Computational Approaches in Physics

Computational Approaches in Physics

Maria Fyta
Institute for Computational Physics, Universität Stuttgart

Morgan & Claypool Publishers

ISBN   978-1-6817-4417-9 (ebook)
ISBN   978-1-6817-4416-2 (print)
ISBN   978-1-6817-4419-3 (mobi)

DOI   10.1088/978-1-6817-4417-9

Version: 20161001

IOP Concise Physics
ISSN 2053-2571 (online)
ISSN 2054-7307 (print)

A Morgan & Claypool publication as part of IOP Concise Physics
Published by Morgan & Claypool Publishers, 40 Oak Drive, San Rafael, CA, 94903 USA

IOP Publishing, Temple Circus, Temple Way, Bristol BS1 6HG, UK

To my high-school and university teachers who taught me how to learn.
To my beloved ones.

Contents

Preface

This book reviews the most common computational schemes used in simulations of physical systems. These range from very accurate *ab initio* techniques up to coarse-grained and mesoscopic schemes. A bottom-up approach is used to present the various simulation methods used in physics, starting from the lower level and most accurate and *ab initio* methods up to mesoscopic particle-based ones. The book outlines the basic theory underlying each technique and its complexity, addresses the computational implications and issues in the implementation, and also presents representative examples. These underline the accuracy and efficiency of the various computational approaches. Illustrating pathways for combining different schemes in order to increase both accuracy and efficiency, while stretching the spatial and temporal scales involved is also discussed. The strengths and deficiencies of the variety of simulational techniques are presented, while many links to the relevant literature are given for further in-depth reading. The present book does not aim to provide an extended view of the methods, rather to touch upon the main general ingredients needed in modeling physical systems and briefly assess the details therein. It is not intended to fully cover the whole spectrum of computational methods and certainly omits many of these. Nonetheless, the aim of this book is to serve as a short guide to computational methods available for physical problems. To supplement this, the most common computational codes, commercial or open source, are listed in the end. It is my hope that the attempt to put together the main elements that constitute computational physics, will be valuable to researchers trying to enter the field or extend their view on computational methods in physics. In this way, they will have strong tools in their hands for tackling a vast variety of problems and open questions in physics.

Acknowledgements

First, I would like to thank the students of my computational courses for helping me improve the content, quality, and level of detail of the teaching material. It was an important element towards the realization of this book. Assistance from D Holder and S Ludwig, both physics students at the University of Stuttgart, in typesetting parts of the text within the framework of a student assignment is greatly acknowledged. This assignment was funded through a teaching grant from the Ministry of Science, Research, and the Arts Baden-Württemberg (MWK) in Germany. Discussions with my former fellow students, colleagues, and graduate students have also been very insightful and deeply influenced the outline of this book. Last, but not least, I'm very much indepted to my advisors and mentors throughout my academic years for their support and everything I have learned from them. This book would not have been possible without the proficiencies I have gained through their guidance.

About the author

Maria Fyta

Maria Fyta studied physics at the University of Crete in Greece, where she obtained her Masters and PhD degree in physics with a specialization in solid state and computational physics under the guidance of P C Kelires. During her graduate studies, she used classical and semi-empirical computational methods, such as Monte Carlo and tight-binding molecular dynamics to study the stability, elastic and mechanical properties of nanostructured carbon. In 2005 she moved to Harvard University as a postdoctoral fellow in the group of E Kaxiras, where she was involved in materials and biophysics related projects, ranging from defects in materials up to the dynamic and electronic behavior of DNA. The computational tools used in these projects were extended to quantum-mechanical schemes, such as techniques based on density functional theory, as well as multiscale schemes based on coupled atomistic and mesoscopic schemes. In 2008 she joined the group of R Netz at the Technical University of Munich, where she was also a Marie Curie fellow. In this group, she turned to atomistic molecular dynamics and the development of atomistic force fields. In 2012 she accepted an offer for a junior professorship from the University of Stuttgart in Germany. Since 2012 she has been affiliated with the Institute for Computational Physics at the University of Stuttgart, where she uses a variety of computational tools to tackle problems in condensed matter physics/materials science and biophysics. Specifically, her field of interests involves nanopores for sensing devices, chemically modified nanostructures, materials with defects, self-assembled monolayers, biofunctionalized materials, as well as the mechanical and electronic properties of biomolecules and their behavior in solutions.

Glossary

BD	Brownian dynamics
BO	Born–Oppenheimer
CC	coupled cluster
CI	configuration interaction
CM	classical mechanics
CG	coarse-graining
CPMD	Car–Parrinello molecular dynamics
DFT	density functional theory
DPD	dissipative particle dynamics
EAM	embedded atom method
FF	force field
GB	generalized Born
HF	Hartree–Fock
$k_B T$	thermal energy
LB	lattice Boltzmann
MC	Monte Carlo
MD	molecular dynamics
MP	Møller–Plesset
MS	multi-scaling
PS	pseudopotentials
QM	quantum mechanics
QMC	quantum Monte Carlo
RNG	random number generator
ROHF	restricted open-shell Hartree–Fock
SCF	self-consistent field
SI	self-interaction
SPH	smoothed particle hydrodynamics
T	temperature
TDDFT	time-dependent density functional theory
UHF	unrestricted Hartree–Fock
XC	exchange-correlation

Chapter 1

Introduction

1.1 Computational physics

Traditionally, physics was divided into two distinct areas, theoretical physics and experimental physics. Theoretical physics solves analytically physical problems, using mathematical tools together with simplified assumptions. Experimental physics observes physical phenomena which happen in the real world. It is desirable that the calculations and observations from both areas meet on common ground. Often though, there are no ways to map the calculations from theoretical physics to the data from experimental physics or experimental data cannot be interpreted at the level of detail inherent in the theoretical models. This occurs because of the complexity of the real world and the simplified assumptions followed in analytical approaches. Nevertheless, sometimes spatial and temporal scales exist at which these two physics areas meet. In these cases, theoretical physics can predict the outcome of experiments, or experimental physics is able to confirm the theory.

Computational physics was introduced in order to cover the gap between the two traditional fields of theoretical and experimental physics and is often based on approaches from theoretical physics. With the aid of modern computing and supercomputers, computational physics can solve highly complex problems and move closer to experimental conditions. In this way, it is possible to promote the physical understanding of experimental observations more efficiently than with only the aid of theoretical physics. Accordingly, theoretical and computational physics together possess strong tools to supplement experimental physics and shed light on physical phenomena in the real physical world.

From the point of view of computational physics, efficient simulations of physical systems involve the strong interplay of three factors, as sketched in figure 1.1. These factors include the systems one aims to simulate and their desired properties, the methodology applied and its accuracy, and the temporal and spatial scales relevant to these systems and their properties. More specifically, it is very important to address the relation of these three factors before designing the computer simulation

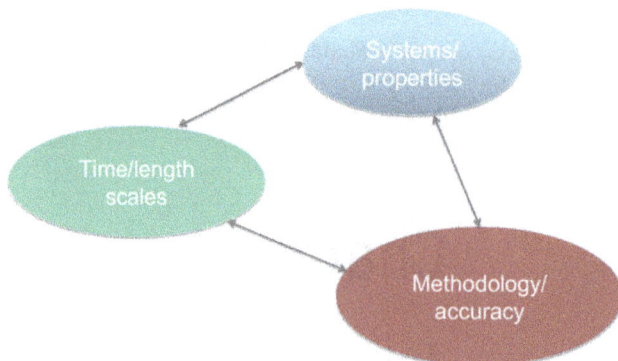

Figure 1.1. The interplay between accuracy, temporal and spatial scales, the characteristics of the modeled systems and their properties probed through computer simulations.

of a physical system. At first sight, one should first choose which are the target properties of the physical system. These will then link to the relevant time and length scales, which in turn can point to the methodology able to extract the desired properties with high accuracy. This process should be followed with care. Choosing, for example, a simulation scheme which cannot efficiently capture the time scales relevant to a desired property of the system can lead to wrong or unphysical results. In principle, the results from such a simulation would be useless.

1.1.1 Length-scales and efficiency

In view of the different temporal and spatial scales involved in a physical system, a number of different simulation methodologies have been developed, denoted by the graph in figure 1.2. Each methodology is based on a description of physics at a specific level and is capable of describing specific properties. Accordingly, each computational scheme can be efficient within a given range of time and length scales. As an example, computational approaches based on quantum mechanics (QM), are quite accurate, can deal with systems of a size up to a few nm and can be simulated for times in the ps range. Due to the high accuracy of the quantum-mechanical simulation methods, these are computationally too demanding and can thus not be applied to systems larger than a few nm. On the other hand, computational models based on a classical description of the system, using for example Newton's equations of motion are efficient for system sizes up to a few μm and times up to msecs. Simulating larger systems becomes computationally very expensive, while going to smaller scales is also not efficient as the classical description is not sufficient for describing these scales and the relevant properties.

In order to make this more intuitive, let's take an example: a physical system is chosen for which the electronic properties need to be analyzed. These are controlled by its electrons (mainly the valence electrons) and their charge distributions within the system. Accordingly, the simulation chosen for this problem needs to be based on a sufficient description of the electronic states. This can be done using quantum-mechanical approaches related to the solution of Schrödinger's equation. This

Hierarchy of scales

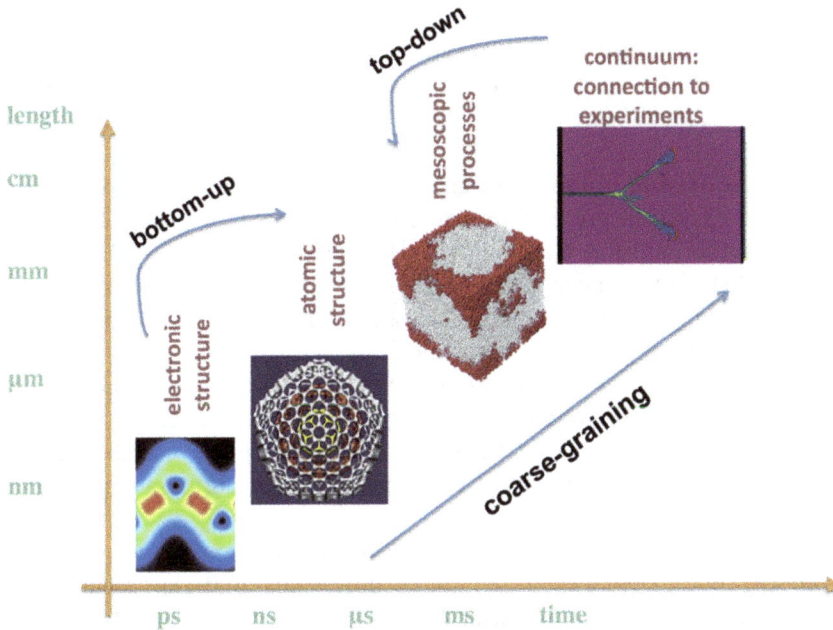

Figure 1.2. The hierarchy of spatial and temporal scales involved in physical systems probed with computer simulations. Parts of this image are reproduced from [1–4] with permission from Cambridge University Press, Chris Ewels (www.ewels.info), the Royal Society of Chemistry, and SIAM, respectively.

equation includes the available electronic states in a system, which increase considerably the larger the system, i.e. the more the atoms and the electronic orbitals composing it. In this respect, such a very accurate scheme involves very many electronic states with increasing system size having a strong impact in solving Schrödinger's equation. The more electrons (i.e. electron states) are involved, the higher the phase space this equation covers leading to highly demanding computations for solving the equation and extracting its eigenvalues to obtain the desired properties. As a consequence, a system with many heavy atoms includes enough electrons to make the solution of Schrödinger's equation through computational means impossible. In this sense, a very accurate quantum-mechanical method can only work at the lower regime of the length versus time graph in figure 1.2.

Essentially, the two lowest regimes in that graph involve the solution of two popular equations: the Schrödinger equation in the very low regime and Newton's equations of motion in the neighboring regime. Although the former equation could be applied to this neighboring regime, it is restricted due to computational reasons, as discussed above. At the same spatial and temporal scale though, Newton's equation of motion could be easily solved using computer simulations. Let's imagine a simple system composed of 10 atoms, each of which includes five electrons. At the quantum-mechanical level, the solution of Schrödinger's equation involves $5 \times 10 = 50$ particles when including all electrons. At the classical level, the solution of

Newton's equation of motion involves 10 particles. Accordingly, at the quantum-mechanical level, solving Schrödinger's equation without any approximation is computationally more difficult than using a classical computational scheme. Nevertheless, in the latter case, properties controlled by the electrons are either approximate or cannot be studied at all. The systems and properties that different computational schemes can describe, as well as the relevant scales, are briefly summarized:

- quantum dynamics and quantum mechanics deal with atoms, nuclei, and electrons, and can provide information on ground/excited states, relaxation, reaction mechanisms, etc on a ps timescale.
- classical statistical mechanics (molecular dynamics, Monte Carlo, force-field methods) model atoms and molecules, and can calculate ensemble averages, dynamics, etc up to ns–ms timescales.
- statistical mechanics models groups of atoms, residues or molecules and can give information on structural homology or similatiry, etc.
- continuum methods deal with the electrical or velocity continuum and can model rheological phenomena, crack propagation, etc on a supra-molecular timescale.
- the kinetic equation on a mesoscopic level can map populations of species in order to obtain population dynamics, flow fields, materials failure, etc. and works in macroscopic time scales.

In principle, in order to achieve a successful and efficient simulation of a physical system, i.e. obtain reliable results with a predictable power, the cooperation of different areas is necessary. The real world can be either observed through an experiment or modeled through computer simulations. The computer simulations involve different levels of accuracy and often depend on parameters obtained from experimental findings. The experimental data are compared with results from simulations or used to test and benchmark simulation methods. On the other hand, when modeling schemes are highly accurate they can lead to predictions at scales not accessible to the experiments. In this way, computer simulations have the power to lead to a better understanding of the physical world.

1.1.2 Approaches and milestones

Since the early (modern) days of computing, a lot of effort was allocated in proposing simulation methods which could describe physical systems within the accuracy and capabilities of the available computational resources. In the early days, simulation methods could be applied to small systems described using stochastic approaches. As computers became faster, more details of a physical system could be incorporated realizing the use of more sophisticated simulations. Over the years, important milestones following the realization of Moore's law [5] could promote more efficient modeling, which in turn better described the real world. In this respect, advances in computer methods in physics have made a direct impact on the cooperation of computational physics and experimental physics, shedding light on a deeper physical understanding of the physical world.

Each simulation scheme developed over the years has made its own important contribution to the field. Nevertheless, a few are the milestones which had a great influence in bringing computational physics to where it is presently. For this, the **Monte Carlo (MC)** method as implemented using the Metropolis algorithm [6, 7] should be acknowledged. This is a stochastic method developed in the 1950s, which was not able to address important atomistic details, but could address important statistical and thermodynamic properties of a physical system. MC is based on statistical physics and the generation of random numbers and can solve problems with a probabilistic interpretation. Approximately 10 years later, the first **molecular dynamics (MD)** studies were able to solve Newton's equations of motion and explain the dynamical properties of a system deterministically [8]. Within MD, the particles can interact through potentials, the accuracy of which controls the efficiency of an MD simulation, hence the trustfulness of the simulated properties. About 15 years after MD, a new computational methodology known as the **density functional theory (DFT)** came to fill the gap at very low scales, at which electrons become important [9]. DFT uses a sophisticated approximation for the Schrödinger equation based on the electron density. Having a satisfying accuracy, it significantly accelerates the simulations and provides an insight into the nanoworld. At this point, the significant contribution of the **Hartree–Fock** theory in the 1930s should be underlined [10, 11]. This theory describes the electronic wavefunctions using a single Slater determinant. A computer implementation of this method accelerates the respective simulations compared to the case where a more extended phase space is considered. Finally, a considerable advancement in computational physics was possible with the **Car–Parrinello molecular dynamics (CPMD)** scheme [12]. CPMD combines MD with DFT, allowing for the inclusion of additional degrees of freedom in the simulations. Specifically, both the electron as well as the atom dynamics are considered. In this way, the application of CPMD has the ability to expand the scales on which quantum mechanical methods are applicable, efficiently add dynamics, and investigate larger physical systems at the same level of accuracy as with DFT.

Often properties of a system at different levels of description are needed, which is usually not possible to achieve with a simulation based on a sole computational method. For example, information on the electronic properties, as well the dynamic and thermodynamic properties of atoms or molecules can be important for a better understanding of a physical system. To this end, different approaches have been developed. The combination of the information obtainable through different computational schemes can basically be done in two ways: (a) through **coarse-graining** (CG) and (b) using **multi-scaling** (MS). Overall, CG or MS is attempted when a simulation can be accelerated without strongly compromising the accuracy of the results. In principle, CG, involves the reduction of the degrees of freedom of a system and is usually done by extracting information from the lower, but most accurate, scales in order to pass the information to the higher scales. MS involves the combination of two or more computational schemes and can follow a bottom-up or a top-down approach as depicted in figure 1.2.

It should have become clear at this point that different computational schemes relate to different degrees of detail, accuracy, computational load, and target properties. A matter of high interest in computer science with a high impact in performing computer simulations in physics is the optimization of the simulation codes. The optimization can be done using parallelization techniques for the codes or computer architectures (e.g. GPUs, CPUs, multi-CPUs, etc) on which the code can be most efficiently ported on. The relevant value is the scaling of the simulation code with the size of the simulated system. This scaling essentially implies the computational load of a specific simulation and the respective computational time needed to perform a simulation. Given that N denotes the size of the system, methods within the scales depicted in figure 1.2 can show a scaling in the range $[N\text{-}N^{10}]$. Note, that N could be either the number of orbitals (in a quantum-mechanical scheme) or the number of atoms (in an atomistic scheme) or the number of the representative particles (at even higher scales). The field of computer simulations in science is continuously evolving, attempting to extend time and length scales and stretch the capabilities of recent supercomputers. Sophisticated schemes for using the computers go beyond what will be reviewed here. An example is the distributed computing project Folding@Home project [13], within which people from any part of the world can provide their computers to band together to the largest supercomputer in the world and extend even more the scales involved in biomolecular simulations.

In the following chapters, a bottom-up approach will be followed to touch upon several of the most common computational methods used to model physical problems. At the lowest level having the highest accuracy, quantum-mechanical simulation methods will be presented, followed by semi-empirical and atomistic schemes moving up to mesoscopic approaches. CG, MS, as well as force-field development will be discussed. This book is intended to present a brief overview of the basic building blocks comprising the most common simulation methods in physics. Short threads to simulations in physics will be given only touching upon the wide possibilities these can offer. Many details will be omitted, but can be found elsewhere [14–19]. The long list of the relevant literature will open up the possibilities for further reading.

1.1.3 Setting-up the simulations

The efficiency of a simulation depends on the system under study, the chosen computational method, the computer architecture, as well as the whole set-up of the simulations. In principle, all factors are interlinked. For example, the simulation scheme is based on the physical problem, and the computer architecture is chosen depending on the simulation code and the desired efficiency/speed. Setting-up the simulations involves roughly the input of the system, the interactions therein, and other relevant simulation parameters.

Most of the simulations in physics involve particles. The position of these particles, whether atoms, molecules, or mesoscopic species are the most common input for the simulations. These particles are typically interacting, and good interaction potentials are necessary for a good description of the system. The positions of these particles can be either obtained using short home-made structure generator codes[1] or other available (commercial or open source) tools – a very few examples of which are Avogadro [20], VirtualNanoLab [21], CRYSTAL [22], SPGEN [23], the DNA server [24] and PHENIX [25]. An alternative is to extract the input data from available data in crystallographic databases [26–28]. Often, the simulation codes provide tools, which assist in generating the system's input. It is, also, essential to use the best potential parameters for the simulated physical system and the choice of the potential needs to be treated with care. A careful choice of the simulation parameters is also essential. These are parameters which control the exact flow and accuracy of the simulation and are related to the specific simulation method. Cut-offs, time-steps, tolerance values, certain conditions (temperature, pressure, etc) are typical simulation parameters. In the end, it is extremely important to be aware of the errors in the simulations. Each simulation method includes an inherent error, but the numerical errors can be estimated through an error analysis, which includes statistical errors and autocorrelation effects [15, 18, 29, 30].

An important step in performing a simulation is the post-processing of the data. Each simulation method can focus on different properties and involves its own way of quantitatively analyzing the simulation data. Often a qualitative visualization of the data can be very insightful. Simulation codes sometimes offer such visualization tools, though it is most common to use the various stand-alone software packages which have been developed for the visualization of scientific data. These can usually do more than just a graphic representation of the particle positions in the simulations. Visualization not only involves the presentation of the particle positions, it can also include additional information linking to the properties of the system. In this way, streamlines can be drawn to show the velocity profile of a solvent, isosurfaces or volume rendering can be used to provide the charge density distribution around atoms, contour lines are able to reveal free energy surfaces, etc. A variety of visualization software packages have been developed and only representative packages are referenced here, such as VMD [31], VESTA [32], Xcrysden [33], GaussView [34] or Xmol [35], etc.

Briefly, when attempting to simulate a physical system, the configurations (particle positions and sometimes their velocities) are needed. As discussed, of high importance are the interactions between the particles, which are either defined *ab initio* or are given in the form of force-fields/potentials. In any case, the chemical identity of the particles is often used as an input. Typically, a 3D computational box (or supercell) is taken which includes all the particles and the size of this box is needed. Note, that 1D or 2D geometries are also preferred when trying to reduce the degrees of freedom, for example in a MC simulation. The strategy in setting-up the

---

[1] These are easy to generate for crystal structures and periodic systems.

simulation of a physical system involves a balanced and careful choice of parameters such as the degrees of freedom to be considered, the interactions/force-fields needed, any sampling and boundary conditions, as well as the environmental conditions (temperature, pressure, external forces) for which relevant schemes need to be accounted for, as will be discussed in the following chapters. A nice review of all the important aspects in molecular simulations can be found elsewhere [36].

References

[1] Kaxiras E 2003 *Atomic and Electronic Structure of Solids* (Cambridge: Cambridge University Press)

[2] http://www.ewels.info (accessed 11 July 2016)

[3] Iype EACC Esteves *et al* 2016 Mesoscopic simulations of hydrophilic cross-linked poly-carbonate polyurethane networks: structure and morphology *Soft Matter* **12** 5029–40

[4] Du Q and Lipton R 2014 Peridynamics, fracture, and nonlocal continuum models *SIAM News* **47** 4(3)

[5] Moore G E *et al* 1998 Cramming more components onto integrated circuits *Proc. IEEE* **86** 82–5

[6] Metropolis M and Ulam S 1949 The Monte Carlo method *J. Am. Stat. Assoc.* **44** 335–41

[7] Metropolis N, Rosenbluth A W, Rosenbluth M N, Teller A H and Teller E 1953 Equation of state calculations by fast computing machines *J. Chem. Phys.* **21** 1087–92

[8] Alder B J and TE W 1959 Studies in molecular dynamics. I general method *J. Chem. Phys.* **31** 459–66

[9] Kohn W and Sham L J 1965 Self-consistent equations including exchange and correlation effects *Phys. Rev.* **140** A1133

[10] Hartree D R *The Wave Mechanics of an Atom With a Non-Coulomb Central Field. Part I. Theory and Methods* vol 24 (Cambridge: Cambridge University Press)

[11] Fock V 1930 Näherungsmethode zur lösung des quantenmechanischen mehrkörperproblems *Zeitschrift für Physik* **61** 126–48

[12] Car R and Parrinello M 1985 Unified approach for molecular dynamics and density-functional theory *Phys. Rev. Lett.* **55** 2471

[13] http://folding.stanford.edu/h (accessed 23 July 2016)

[14] Frenkel D and Smit B 2001 *Understanding Molecular Simulation: From Algorithms to Applications* vol 1 (Orlando, FL; Academic)

[15] Leach a R 2001 *Molecular Modelling: Principles and Applications* (Englewood Cliffs, NJ: Pearson)

[16] Rapaport D C 2004 *The Art of Molecular Dynamics Simulation* (Cambridge: Cambridge University Press)

[17] Thijssen J 2007 *Computational Physics* (Cambridge: Cambridge University Press)

[18] Steinhauser M O 2008 *Computational Multiscale Modeling of Fluids and Solids* (Berlin: Springer)

[19] Scherer P O J 2008 *Computational Physics* (Berlin: Springer)

[20] http://avogadro.cc/wiki/Main_Page (accessed 2 February 2015)

[21] http://quantumwise.com/products/vnl (accessed 2 February 2015)

[22] http://www.crystal.unito.it (accessed 15 July 2016)

[23] http://xtalopt.openmolecules.net/spgGen/spgGen.html (accessed 15 July 2016)

[24] http://structure.usc.edu/make-na/ (accessed 15 July 2016)

[25] https://www.phenix-online.org (accessed 15 July 2016)

[26] http://www.crystallography.net/cod (accessed 15 July 2016)

[27] https://cds.dl.ac.uk/cds/datasets/crys/icsd/llicsd.html (accessed 15 July 2016)

[28] http://www.rcsb.org (accessed 15 July 2016)

[29] Allen M P and Tildesley D J 1989 *Computer Simulation of Liquids* (Oxford: Oxford University Press)

[30] Hamming R 2012 *Numerical Methods for Scientists and Engineers* (New York: Dover)

[31] http://www.ks.uiuc.edu/Research/vmd (accessed 2 February 2015)

[32] http://jp-minerals.org/vesta/en/ (accessed 2 February 2015)

[33] http://www.xcrysden.org/ (accessed 2 February 2015)

[34] http://www.gaussian.com/g_prod/gv5.htm (accessed 15 July 2016)

[35] http://www.molecular-explorer.com (accessed 15 July 2016)

[36] van Gunsteren W F *et al* 2006 Biomolecular modeling: goals, problems, perspectives *Angew. Chem., Int. Ed. Engl.* **45** 4064–92

Computational Approaches in Physics

Maria Fyta

Chapter 2

Quantum-mechanical methods

2.1 General remarks

Quantum-mechanical (QM) methods, often named as 'electronic structure methods', involve the shortest temporal and smallest spatial scales depicted in figure 1.2. QM schemes can be applied to relatively small systems, but can have a high level of accuracy and typically do not depend on empirical parameters. There are different levels of QM methods depending on the degree of accuracy in the description of the system. All, though, are based on solving the Schrödinger equation in its general time-dependent form:

$$i\hbar\frac{\partial}{\partial t}\Psi(\vec{r}, t) = \hat{H}\Psi(\vec{r}, t) \qquad (2.1)$$

where $\Psi(\vec{r}, t)$ is the total wavefunction of the QM system, while $\vec{r}$, t are the position and time variables. $\hat{H}$ describes the Hamiltonian of the system. A general form of this Hamiltonian for a many-body system including atoms, their nuclei and electrons, would read

$$H = -\sum_i \frac{\hbar^2 \nabla_r^2}{2m_e} - \sum_I \frac{\hbar^2 \nabla_R^2}{2M_I} - \sum_{i,I} \frac{Z_I e^2}{\left|\vec{R}_I - \vec{r}_i\right|} + \frac{1}{2}\sum_{i \neq j} \frac{e^2}{\left|\vec{r}_i - \vec{r}_j\right|^2} + \frac{1}{2}\sum_{I \neq J} \frac{Z_I Z_J e^2}{\left|\vec{R}_I - \vec{R}_J\right|^2} \quad (2.2)$$

In this equation, the indices i and I run over the total number of electrons and nuclei with mass m_e and M_I, respectively. The positions of the electrons and the atoms in space are given through $\vec{r}_i$ and $\vec{R}_I$. The atomic number of the nuclei is Z_I and e is the electron charge. The first two terms in this equation are the kinetic energy terms of the electrons and the nuclei, respectively. All other terms correspond to the electron–nuclei, electron–electron, and nuclei–nuclei Coulomb interactions, respectively.

For a many-body system, solving equation (2.1) with a Hamiltonian from equation (2.2) becomes extremely difficult or even impossible as the system size increases. Accordingly, one of the main approximations taken is the **Born–Oppenheimer (BO) approximation**. Within this approximation, ions move slowly in space, and electrons respond instantaneously to any ionic motion. Because of this, the wavefunction Ψ explicitly depends only on the electronic degrees of freedom. Accordingly, the potential between the nuclei and the kinetic energy of the nuclei are taken as constants as far as the electronic degrees of freedom are concerned. Equation (2.2) is then simplified to

$$H = -\sum_i \frac{\hbar^2 \nabla_r^2}{2m_e} - \sum_i V_{\text{ion}}(\vec{r_i}) + \frac{e^2}{2} \sum_{i \neq j} \frac{1}{|\vec{r_i} - \vec{r_j}|^2} \qquad (2.3)$$

where the total external potential experienced by an electron due to the presence of ions is defined as $V_{\text{ion}}(\vec{r_i}) = -\sum_{i,I} \frac{z_I e^2}{|\vec{R_I} - \vec{r_i}|}$.

Before moving to the description of the QM simulation methods, a few important ingredients of these need to be presented. The **exchange** and **correlation** interactions between electrons are of high importance in determining the energy of a QM system [1, 2]. The term **exchange** denotes the interactions between electrons with parallel spins, while the term **correlation** corresponds to the Coulomb interaction between electrons. Often the respective interactions are gathered together in the **exchange-correlation** term in the equations for the electrons. Depending on the degree of accuracy, QM approaches may or may not include the exchange and/or correlation interactions. An additional aspect is the **self-interaction (SI)**. term, which is the contribution to the energy of a particle due to the interactions of the particle with the system it is part of. A known issue in QM calculations is **spin contamination**, which involves the artificial mixing of different electronic spin-states and can occur when the spatial parts of α and β spin-orbitals are allowed to differ [3]. Wavefunctions with a high degree of spin contamination are undesirable, as these are not eigenfunctions of the total spin-square operator. Depending on the exact QM method, this issue can be resolved in different ways [4, 5]. Overall, the QM simulation methods mainly differ on the approximations taken to solve equation (2.1). Most of the QM methods involve static ground state properties. It will be seen though, that the temporal variation can also be accounted for and probe also dynamical aspects. In the following sections, the most important aspects of the QM simulation methods used in physics will be underlined. Additional schemes do exist, but are not presented here. Note that the discussion on the quantum MC methods, which is also based on solving the Schrödinger equation is left for section 6.3. The reader is directed to the literature for more extensive presentations [6–9].

2.1.1 Two descriptions for the electronic structure methods

In essence, the QM schemes can all be tracked down to the calculation of the total electronic energy of the system. For this, two are the different descriptions of a QM many-body system on which the QM simulation methods are based: (a) the **wavefunction** world (introduced by Pople [7]) in which the total electron energy E_e of the system is defined as: $E_e = E_e[\psi_e(\vec{r_1}, \ldots, \vec{r_N})] = \langle \psi_e | H | \psi_e \rangle$ where $\psi_e(\vec{r_1}, \ldots, \vec{r_N})$ is the electronic wavefunction for the electrons at positions $\vec{r_1}, \ldots, \vec{r_N}$ and the many-body problem is reduced to a $3N$-dimensional variational problem based on a set of one-particle equations. Using the variational principle, the optimized electron energy of the QM system, E_e^{opt}, can be obtained through the minimization of the electronic energy with respect to the electronic wavefunction. (b) The **density** world (introduced by Kohn, Hohenberg, and Sham [10, 11]) in which the total electron energy E_e of the system is defined as $E_e = E_e[n(\vec{r})]$ where $n(\vec{r})$ is the electron density of the many-body system and the theory is reduced to a 3-dimensional variational problem. Using again the variational principle, E_e^{opt} can be obtained through the minimization of the electronic energy with respect to the electron density. Both descriptions, though, had a great impact in developing QM computational schemes efficiently describing the real world.

2.2 The Hartree–Fock method

In essence, the basic contribution to the modern QM simulation techniques originates from the Hartree–Fock (HF) methods [12, 13]. The HF theory uses the description of the 'wavefunction world' of the one-electron picture. HF assumes that each electron is moving independently of the others in the field created by the fixed nuclei (BO approximation) and the mean-field of the other electrons. The HF method is an approximate solution to the Schrödinger equation, which requires that the final field computed from the charge distribution is 'self-consistent' with the assumed initial field (self-consistent field method SCF, Hartree 1927).

The Hartree approximation assumes that the electrons are non-interacting particles. The electron exchange and correlation are taken into account in an average way through the mean field approximation for the electron–electron interactions, which is actually a severe simplification. The next level of sophistication would be the inclusion of the fermionic nature of the electrons, which is done within the HF theory. Accordingly, the exchange property of the electrons is accounted for in HF, but not the electron correlation. The electronic wavefunction is mapped through a single Slater determinant. Assuming that $\phi_i(\vec{r}_i)$ are the normalized single-particle states with i the index running over the single particles, i.e. the electrons, the following Slater determinant can then describe the many-body wavefunction Ψ^{HF} within the HF framework. Ψ^{HF} is antisymmetric as the sign of the Slater determinant changes sign when the coordinates of two electrons are interchanged. The spin degrees of freedom are included by considering both electrons with up and down spins at position $\vec{r}$. For N electrons, the expression for the many-body wavefunction is then $\Psi^{\text{HF}}(\{\vec{r}_i\}) = \frac{1}{\sqrt{N!}}\det\{\phi_i(\vec{r}_i)\}$: the total energy of the system is then the expectation value of the Hamiltonian of the many-body systems with respect to Ψ^{HF}.

$$
\begin{aligned}
E^{\text{HF}} &= \left\langle \Psi^{\text{HF}} \middle| H \middle| \Psi^{\text{HF}} \right\rangle \\
&= \sum_i \left\langle \phi_i \middle| -\frac{\hbar^2 \nabla_r^2}{2m_e} + V_{\text{ion}}(\vec{r}) \middle| \phi_i \right\rangle \\
&\quad + \frac{e^2}{2} \sum_{i \neq j} \left\langle \phi_i \phi_j \middle| \frac{1}{|\vec{r} - \vec{r}'|} \middle| \phi_i \phi_j \right\rangle
\end{aligned}
\tag{2.4}
$$

In this equation, the first summation includes the kinetic energy and the ionic potential, i.e. the electron–nuclei interactions. The second summation accounts for the electron–electron interactions. Variational arguments lead to the single-particle HF equation [12, 14]:

$$\varepsilon_i \phi_i(\vec{r}) = \left[-\frac{\hbar^2 \nabla_r^2}{2m_e} + V_{\text{ion}}(\vec{r}) + V_i^H(\vec{r}) \right] \phi_i(\vec{r})$$

$$- e^2 \delta_{s_i s_j} \sum_{j \neq i} \left\langle \phi_j \left| \frac{1}{|\vec{r} - \vec{r}'|} \right| \phi_i \right\rangle \phi_j(\vec{r})$$

(2.5)

The second term is the exchange term and it is the additional term compared to the respective Hartree single-particle equation. That is the 'exchange' term between the electrons. $V_i^H(\vec{r})$ is the Hartree potential, which is different for each electron and includes only the repulsion from other electrons. $s_{i/j}$ is a label for the spin of particles i, j. HF simulations involve the self-consistent solution of equation (2.5) according to the iterative algorithm in figure 2.1. An initial wavefunction $\phi_i^{\text{in}}(\vec{r})$ is chosen. Based on this, the HF equations are iteratively solved for a new wave-function $\phi_i^{\text{out}}(\vec{r})$. The agreement of $\phi_i^{\text{in}}(\vec{r})$ and $\phi_i^{\text{out}}(\vec{r})$ within a certain tolerance defines the end of the simulation.

Many of the exact mathematical details of the HF method have been omitted here, but an extensive analysis can be found elsewhere [8, 9, 15]. Note, that E^{HF} is the lowest possible energy neglecting correlation. Within HF all SIs are exactly canceled by the corresponding exchange SIs. Regarding spin contamination mentioned earlier, HF gives rise to different equations for the α and β orbitals and two approaches can be taken. (i) The restricted open-shell Hartree–Fock (ROHF) [16]: enforce a double occupation of the lowest orbitals by constraining the spatial distributions of α and β to be the same. (ii) unrestricted Hartree–Fock (UHF) [17]: permit a complete variational freedom. Overall, HF uses the BO

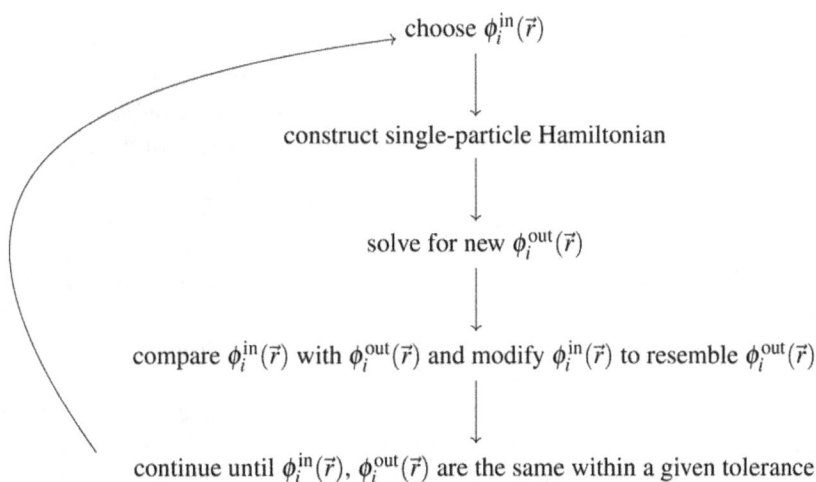

choose $\phi_i^{\text{in}}(\vec{r})$

construct single-particle Hamiltonian

solve for new $\phi_i^{\text{out}}(\vec{r})$

compare $\phi_i^{\text{in}}(\vec{r})$ with $\phi_i^{\text{out}}(\vec{r})$ and modify $\phi_i^{\text{in}}(\vec{r})$ to resemble $\phi_i^{\text{out}}(\vec{r})$

continue until $\phi_i^{\text{in}}(\vec{r})$, $\phi_i^{\text{out}}(\vec{r})$ are the same within a given tolerance

Figure 2.1. The Hartree–Fock SCF algorithm followed in simulations.

approximation, does not include relativistic effects and is a variational solution based on a linear combination of a finite number of basis functions described by a single Slater determinant. The mean-field approximation within HF does not include correlation for electrons with opposite spin, but only for electrons with parallel spins. Depending on the physical problems, these approximations are not always efficient. However, with these the HF method reduces the phase space scanned in the simulations making the modeling of many physical systems possible. The method, however, is expensive and can treat only small systems of the order of 10^2 atoms.

2.3 Post HF schemes

The HF method has brought a significant advancement in the field of simulation methods in physics. However, the simplifications HF assumes, specifically in terms of electron correlation are often too crude and cannot describe with high fidelity a physical problem. In order to include features, like the full correlation, previously omitted and improve the HF scheme, other computational methods known as *post HF methods* have been developed. These are more accurate, but also computationally more expensive than HF, being able to model only systems of a very small size. In the following, the basic concepts behind some of the common post HF approaches are reviewed.

2.3.1 Coupled cluster

The coupled cluster (CC) scheme constructs multi-electron wavefunctions based on HF using an exponential cluster operator, $\hat{T}$, which accounts for the electron correlation [18–20]. Within CC, the total wavefunction, $|\Psi\rangle$, of the system is described through:

$$|\Psi\rangle = e^{\hat{T}}|\phi_0\rangle$$

$|\phi_0\rangle$ is a Slater determinant usually constructed from the Hartree–Fock wavefunctions. The cluster operator is an excitation operator which, when acting on $|\phi_0\rangle$, produces a linear combination of excited Slater determinants. $\hat{T}$ converts the reference $|\phi_0\rangle$ into a linear combination of single- and double-excited Slater determinants. When single(S), double(D), and triple(T) excitations are taken into account, then the cluster operator can be written as:

$$\hat{T} = \hat{T_1} + \hat{T_2} + \hat{T_3} + \cdots$$

The separate operators can be written in the second quantization formalism and include creation and annihilation operators for occupied and unoccupied orbitals, respectively. The corrections of a given type (S,D,T,...) to infinite order denote the CC approaches of a different accuracy (CCS, CCSD, CCSDT, ...). In practice the series are finite due to the finite number of orbitals and excitations. CCSD(T) includes single, doubles fully, but triplets are calculated with perturbation theory. CCSD(T) is more accurate and expensive than CCSD. The CC scheme is non-variational, size-extensive and size-consistent.

2.3.2 Møller–Plesset perturbation theory

The Møller–Plesset (MP) approach adds correlation effects to improve HF [21] based on the Rayleigh–Schrödinger perturbation theory [22, 23]. What this scheme, roughly, does is to take a perturbed Hamiltonian $\hat{H}$ which differs slightly from the unperturbed reference Hamiltonian $\hat{H}_0$ with a perturbation $\hat{V}$. Within the MP framework, the unperturbed Hamiltonian is taken as the sum of single particle Fock operators ($\hat{F}_i$), which include the electron–electron repulsion in an average way. The Fock operator approximates the single-electron energy operator of a given quantum system with a given set of basis vectors and is an approximation to the true Hamiltonian [24]:

$$\hat{F}_i = \hat{h}_i + \sum_{j=1}^{N}\left(\hat{J}_i - \hat{K}_i\right) \tag{2.6}$$

with $\hat{h}_i$ the one-electron Hamiltonian, $\hat{J}_i$ the Coulomb operator, $\hat{K}_i$ the exchange operator, and i runs over all electrons. N is the total number of occupied orbitals. Then the unperturbed Hamiltonian can be written as:

$$\hat{H}_0 = \sum_{i=1}^{N}\hat{F}_i = \sum_{i=1}^{N}\hat{h}_i + \sum_{i=1}^{N}\sum_{j=1}^{N}\left(\hat{J}_{ij} - \hat{K}_{ij}\right) = \sum_{i=1}^{N}\hat{h}_i + \sum_{i,j}^{N}\hat{V}_{ij}^{\mathrm{HF}} \tag{2.7}$$

The difference between the full Hamiltonian and the Fock operator is then:

$$\hat{V} = \hat{H} - \hat{H}_0 = \sum_{i<j}^{N}\hat{V}_{ij} - \sum_{i,j}^{N}\hat{V}_{ij}^{\mathrm{HF}} \tag{2.8}$$

which is the difference between the instantaneous and average electron–electron interaction. The zero-order wavefunction is the HF determinant and the zero-order energy (E_{MP0}) is the sum of molecular orbital energies:

$$E_{\mathrm{MP0}} = \sum_{i=1}^{N}\left\langle \phi_i|\hat{F}_i|\phi_i\right\rangle = \sum_{i=1}^{N}\varepsilon_i^{\mathrm{HF}} \tag{2.9}$$

The HF orbital basis $\{\phi_i\}$ constructs the Slater determinant many-electron wavefunction ϕ_0. The MP perturbation theory observes that all Slater determinants formed by exciting electrons from occupied to virtual orbitals are also eigenfunctions of $\hat{H}_0$ with an eigenvalue equal to the sum of the one-electron energies of the occupied spin orbitals. At zero-order, double counting of the electron–electron repulsion is being corrected. The first-order energy correction (MP1[1]) is the average of the perturbation operator over the zero-order wavefunction:

$$E_{\mathrm{MP1}} = \left\langle \phi_0|\hat{H}|\phi_0\right\rangle \tag{2.10}$$

The sum of E_{MP0} and E_{MP1} is exactly the HF energy $E_{\mathrm{MP0}} + E_{\mathrm{MP1}} = E_{\mathrm{HF}}$.

---

[1] Note, that E(MPn) indicates the correction at order n, and $E_{\mathrm{MP}n}$ indicates a total energy up to order n.

In MP, the electron correlation starts at order two with this choice of H_0 and involves the sum over double excited Slater determinants (single excited determinants give no contribution to energy). These can be generated by promoting two electrons from occupied orbitals ϕ_a, ϕ_b (with energies ε_a, ε_b) to virtual orbitals ϕ_r, ϕ_s (with energies ε_r, ε_s). The second MP correction is expressed as:

$$E(\text{MP2}) = \sum_{a<b}^{\text{occ}} \sum_{r<s}^{\text{vir}} \frac{\left[\left\langle \phi_a\phi_b |\hat{V}| \phi_r\phi_s \right\rangle - \left\langle \phi_a\phi_b |\hat{V}| \phi_r\phi_s \right\rangle\right]^2}{\varepsilon_a + \varepsilon_b - \varepsilon_r - \varepsilon_s} \tag{2.11}$$

MP2 accounts for ~80–90% of the correlation energy and is the cheapest method for including electron correlation. Typically, MP2, MP3, and MP4 are used. A fifth-order correction (MP5) is too expensive. Note, that no monotonic convergence is observed by moving through HF, MP2, MP3, MP4 and increasing the accuracy. It can be observed that the HF and MP2 results differ, with the MP3 results moving back to HF, and MP4 moving away again. In principle, experience shows that for 'well behaved systems', the correct answer lies between MP3 and MP4. At the MP2 level, approximately 100–150 basis functions can be calculated at a cost similar to HF calculations. This holds when the zero-order (HF) wavefunction is a reasonable approximation to the real wavefunction and the perturbation operator is small. Overall, the convergence of MPn simulations depends on the size of basis set and can be slow, rapid, oscillatory, or even non-existent.

2.3.3 Configuration interaction

Configuration Interaction (CI) improves the HF method by increasing the phase space of the system. This is being done by replacing the single Slater determinant of the HF theory to a set of many Slater determinants through the matrix reformulation of the Schrödinger equation. The term 'configuration' denotes the linear combination of Slater determinants for the wavefunction, while 'interaction' states the mixing of different electronic states. Within CI, non-relativistic multi-electron systems in the Born–Oppenheimer approximation can be treated, but relativistic effects such as spin–orbit coupling can also be described.

The many-body wavefunction for the electronic ground state within the CI framework can be written in terms of 'excitations' from the HF 'reference' Slater determinant using N-electron basis functions $|\psi_i\rangle$:

$$|\Psi_0\rangle = c_0|\Phi_0\rangle + \sum_{ra} c_a^r |\Phi_a^r\rangle + \sum_{a<b,r<s} c_{ab}^{rs} |\Phi_{ab}^{rs}\rangle + \cdots = \sum_{i=0} c_{0i}^r |\Phi_i^r\rangle \tag{2.12}$$

The $|\Phi_a^r\rangle$ Slater determinant is formed by replacing the spin-orbital a in $|\Phi_0\rangle$ with the spin-orbital r. The set of $\{|\Phi_a^r\rangle\}$ are expanded based on the single-particle basis functions $\{\psi_i\}$. Every N-electron Slater determinant can be described by a set of N spin orbitals from which it is formed. This set of orbital occupancies is a **configuration**. A full CI calculation involves the use of a set of one-particle functions $\{\psi_i\}$ and all possible N-electron basis functions $\{|\Phi_i\rangle\}$. The computational cost can be reduced by reducing the CI space through truncation of the CI expansion according to the excitation levels relative to the reference HF state. Commonly, included are only those N-electron basis functions that represent double excitations, a method known as CID. CISD involves single or double excitations. Other methods, like the quadratic CI introduce additional terms, which are quadratic in the configuration coefficients and ensure size consistency in the resulting energy [25].

The total expectation value of the system based on the expression for the CI many-body wave function in equation (2.12) and the Hamiltonian H can be calculated through $E = \frac{\langle \Psi_0 | H | \Psi_0 \rangle}{\langle \Psi_0 \Psi_0 \rangle}$. In CI, a basis $\{|\Phi_i\rangle\}$ needs to be taken in order to solve the matrix eigenvalue problem: $HC = ECI$ and obtain the ground state energy E and the coefficients $\{c_{0i}\}$ in equation (2.12). The hermitian Hamiltonian matrix H can be written as blocks of single ($|S\rangle$), double ($|D\rangle$), triple ($|T\rangle$) and quadruple ($|Q\rangle$) excited determinants

$$
H = \begin{array}{c} \langle \psi_0 | \\ \langle S | \\ \langle D | \\ \langle T | \\ \langle Q | \\ \vdots \end{array} \left(\begin{array}{ccccccc} \langle \Phi_0 | H | \Phi_0 \rangle & \cdots & & & & & \cdots \\ 0 & \langle S|H|S \rangle & \cdots & & & & \cdots \\ \langle D|H|\Phi_0 \rangle & \langle D|H|S \rangle & \langle D|H|D \rangle & \cdots & & & \cdots \\ 0 & \langle T|H|S \rangle & \langle T|H|D \rangle & \langle T|H|T \rangle & \cdots & & \cdots \\ 0 & 0 & \langle Q|H|D \rangle & \langle Q|H|T \rangle & \langle Q|H|Q \rangle & \cdots \\ \vdots & \vdots & \vdots & \vdots & \vdots & \vdots & \vdots \end{array} \right) \quad (2.13)
$$

H is symmetric if only real orbitals are used. In CI only doubles interact with the HF reference, thus double excitations are expected to make the largest contribution to the CI wavefunctions beyond the reference state. Singles, triples, and higher excitations can still be part of CI wavefunction with non-zero coefficients, as the states can mix with doubles directly or indirectly. Singles contribute very little to the energy compared to doubles, but are important in describing one-electron properties and are computationally cheap. For single and double excitations about 95% of the ground state correlation energy in small molecules is captured. CI is very demanding and can be applied to systems of a few atoms.

2.4 Density functional theory (DFT)

The HF and post HF methods are all based on attempts to use sufficient approximations to the many-body wavefunctions in order to solve the Schrödinger equation. In 1964 Hohenberg and Kohn proved that the exact ground state electron density of a system **uniquely** specifies the acting single-electron potential [10]. This density controls the Hamiltonian of the system and thereby all properties of the ground state. This can be interpreted with the existence of an energy functional $E[\rho]$, which gives the exact ground state energy for the exact ground state density. The energy is minimized for this exact density. Kohn and Sham [11] presented a variational procedure to simply implement the DFT[2] in computational schemes. DFT significantly accelerated the calculations for quantum mechanical systems leading to larger scale simulations.

---

[2] Also known as the Hohenberg–Kohn–Sham theory.

The basic concept of DFT is that instead of the many-body wavefunction, $\Psi(\{\vec{r}_i\})$, the total density of the electrons in the system, $n(\vec{r})$ is considered. It is a conceptually simple idea without a need to define $\Psi(\{\vec{r}_i\})$. In principle, DFT constructs single-particle equations exactly and then introduces approximations. Assuming $\Psi(\vec{r}) = \Psi(\vec{r}_1, \vec{r}_2, ..., \vec{r}_N)$ to be the total wavefunction of an electronic system, the electronic density of the ground state can be defined as

$$n(\vec{r}) = N \int \psi^*(\vec{r}_1, ..., \vec{r}_N)\Psi(\vec{r}_1, ..., \vec{r}_N)d\vec{r}_2...d\vec{r}_N \qquad (2.14)$$

where $\vec{r}_i$ is the position of electron i. The Hamiltonian of the system depends on the universal operators for any N-electron system, the kinetic energy T and the electron–electron interaction W, and on the external potential V, which depends on the system. A *universal functional of density* could then be written:

$$F[n(\vec{r})] = \langle \Psi(n)|(T + W)|\Psi(n)\rangle = \text{const.} \qquad (2.15)$$

and the total energy of the system can also be expressed as a functional through:

$$E[n(\vec{r})] = \langle \Psi \mid H|\Psi \rangle = F[n(\vec{r})] + \int V(\vec{r})n(\vec{r})d\vec{r} \qquad (2.16)$$

By expressing the functional based on one- and two-particle densities and the basis functions ($\phi_i(\vec{r})$), it can be further expressed as:

$$F[n(\vec{r})] = T^S[n(\vec{r})] + \frac{e^2}{2} \int\int \frac{n(\vec{r})n(\vec{r}')}{|\vec{r} - \vec{r}'|}d\vec{r}d\vec{r}' + E^{XC}[n(\vec{r})] \qquad (2.17)$$

The term $E^{XC}[n(\vec{r})]$ includes all exchange and correlation effects of the many-body character of the true electron system. The use of the variational principle for the density and Langrange multipliers leads to the single-particle equations for the potential energy:

$$\left[-\frac{\hbar^2}{2m_e}\nabla^2 + V_{\text{eff}}\left(\vec{r}, n(\vec{r})\right)\right]\phi_i(\vec{r}) = \varepsilon_i\phi_i(\vec{r}) \qquad \text{Kohn–Sham equation} \qquad (2.18)$$

The effective potential is also a function of the electron density and depends on all single-particle states:

$$V_{\text{eff}}\left(\vec{r}, n(\vec{r})\right) = V(\vec{r}) + e^2 \int \frac{n(\vec{r}')}{|\vec{r} - \vec{r}'|}d\vec{r}' + \frac{\delta E^{XC}\left[n(\vec{r})\right]}{\delta n(\vec{r})} \qquad (2.19)$$

The first term in equation (2.19) is the external potential due to the ions, the second term is the kinetic energy, and the last term is the variational functional derivative of the exchange–correlation interaction. The single-particle orbitals $\phi_i(\vec{r})$ are known as the Kohn–Sham orbitals, which represent states of fictitious (non-interacting) fermionic particles with the same density as that of the real electrons. The solution of the system can be computed by solving the Kohn–Sham equations in equation (2.18) by iteration until self-consistency is reached. A complex issue within DFT is the exact form of $E^{XC}[n(\vec{r})]$, which is unknown. Possible approaches to the

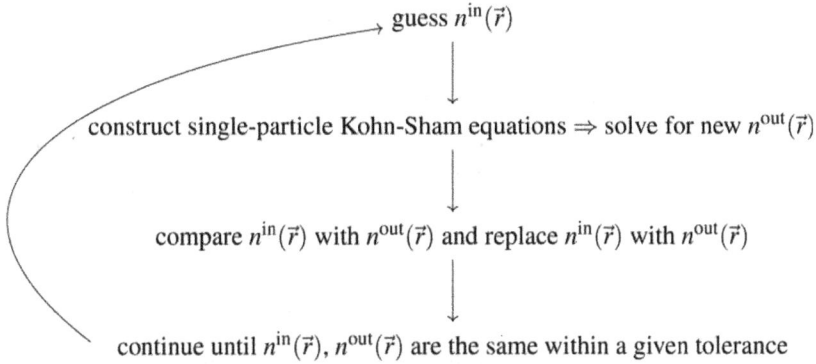

Figure 2.2. The algorithm followed in DFT simulations.

exchange-correlation functional in order to capture the many-body effects in an efficient way will be discussed in section 2.4.1. In most of the commonly used models, the exchange-correlation functional is decomposed into a correlation and an exchange term, respectively:

$$E^{XC}[n(\vec{r})] = \int \left\{ \varepsilon^x[n(\vec{r})] + \varepsilon^{cor}[n(\vec{r})] \right\} n(\vec{r}) \mathrm{d}\vec{r} \qquad (2.20)$$

The correlation energy affects both kinetic and potential energies and is more complicated to calculate than the exchange energy. Correlation, though, is more important for electrons of opposite spin, since electrons of the same spin are automatically kept apart by the Pauli exclusion principle. Note, that for the ground state the correlation energy is always negative. Excited states are not well captured through the typical DFT. Additional details on deriving the Kohn–Sham equations can be found elsewhere [8, 26]. Regarding spin contamination, often DFT calculates spin-contamination using the KS orbitals as if they were HF orbitals, which is not always correct [27].

Algorithm. A DFT computer algorithm is very similar to the one for solving the HF equations sketched in figure 2.2. A guess for an initial electron density $n^{in}(\vec{r})$ is made and followed by the solution of the respective Kohn–Sham equations. The solution will lead to a new electron density $n^{out}(\vec{r})$, which is compared to the initial guess and is repeated for each SCF step. This procedure is being done iteratively until convergence of the input and output electron density within a given tolerance is reached.

2.4.1 Exchange and correlation functionals

The art in DFT is the choice of the exchange correlation (XC) functional in equation (2.20). A number of different XC functionals have been proposed in the past. It is not possible to review all of these here. The description will rather be restricted to the ones most commonly used in physics. In general, the functional should be non-local, i.e. simultaneously dependent on two positions in space, $\vec{r}$ and $\vec{r}'$, because exchange and correlation effects are inherently non-local in an interacting electron system. Nonetheless, developing non-local exchange-correlation functionals is not an easy task. In the following XC, X, and C denote exchange-correlation, exchange, and correlation, respectively.

Local (spin) density approximation—L(S)DA. The local density approximation (LDA) [11] is the simplest XC functional. In the limit of the homogeneous electron gas, effects of exchange and correlation are local. Within LDA, the exchange-correlation energy is an integral over the whole space with the exchange-correlation energy at each point assumed to be the same as in a homogeneous gas of that density. A generalization of LDA is LSDA, with S standing for the spins. The expression for the LSDA XC functional will be:

$$E_{\text{LSDA}}^{\text{XC}}[n^\uparrow, n^\downarrow] = \int n(\vec{r})\varepsilon_{\text{hom}}^{\text{XC}}\big(n^\uparrow(\vec{r}), n^\downarrow(\vec{r})\big)\mathrm{d}r \tag{2.21}$$

where $n^\uparrow$, $n^\downarrow$ are the densities of the electrons with up and down spins, respectively, $n(\vec{r})$ is the total electronic density at position $\vec{r}$, and hom stands for the homogeneous unpolarized electron gas. L(S)DA is expected to work better for solids close to a homogeneous gas. A known complication of LDA is the spurious self-interaction term, which is negligible in the homogeneous gas, but large in confined systems. The lattice constants within LDA are found 1–3% smaller than experimental values. Cohesive energies correspond to systems 5–20% too strongly bound, while the bulk moduli are 5–20% larger than in experiments. The largest error up to 40% lower than expected can be found in the electronic band gaps.

Generalized-gradient approximation (GGA). GGA [28] is typically an improvement over LSDA. It depends on the electron density and the magnitude of the gradient of this density $|\nabla n|$. A generalized form of GGA for a spin polarized system is expressed as:

$$E_{\text{GGA}}^{\text{XC}}[n^\uparrow, n^\downarrow] = \int n(\vec{r})\varepsilon^{\text{XC}}\big(n^\uparrow, n^\downarrow, |\nabla n^\uparrow|, |\nabla n^\downarrow|\big)\mathrm{d}r \tag{2.22}$$

The up and down arrows link to the expressions for the up and down spins, respectively. The most common forms of $F^{\text{X}}(n, |\nabla n|)$ are those proposed by Becker (B88) [29], Perdew–Wang (PW91) [30], and Perdew–Burke–Enzernhof (PBE) [31]. GGA accounts for a higher complexity compared to LDA, but is not always more efficient. GGA can lead to worse results compared to LDA, because gradients in real materials are so large that a low-order expansion of exchange and correlation breaks down. GGA functionals improve the calculation of the cohesive energies compared to LDA, but are not always better in obtaining lattice parameters. Although LDA could be sufficient for covalently bonded systems where up to d orbitals are involved, the use of GGA is important for magnetic and metallic systems.

Other approximations to XC. The description of the XC functionals within DFT is a very active field of research and the existing functionals are being improved in order to better capture more properties of the real systems. Non-local density formulations are attempted through the average density approximation (ADA) and weighted density approximation (WDA) [32]. Other approaches involve the orbital-dependent functionals, $E^{\text{XC}}[\{\psi_i\}]$, which are the self-interaction corrections (SIC) (correcting for unphysical self-interaction) [33], the LDA+U (in which the LDA or GGA are coupled to additional orbital-dependent interactions for highly localized atomic-like orbitals) [34], the optimized effective potential (OEP) [35, 36], the exact exchange (EXX) (using the HF exchange in terms of orbitals) [12], the Slater local approximation [37], the meta-generalized-gradient approximation (in which a curvature of density is introduced), the GW approximation (using Green's functions to estimate the self-energy) [38] or hybrid functionals, etc. The hybrid functionals include the exact exchange from HF combined with *ab initio* or empirical exchange and correlation terms. Examples of these are the B3LYP [39], HSE [40], and meta-hybrid GGA [41]. The B3LYP XC functional, as an example is

$$E_{B3LYP}^{XC} = E_{LDA}^{XC} + a_0\left(E_{HF}^{X} - c\right) + a_X\left(E_{GGA}^{X} - E_{LDA}^{X}\right) + a_C\left(E_{GGA}^{C} - E_{LDA}^{C}\right) \quad (2.23)$$

with $a_0 = 0.20$, $a_X = 0.72$, and $a_C = 0.81$. E_j^i corresponds to the exchange-correlation, exchange, and correlation ($i = XC, X, C$) terms in other approximations ($j = LDA, HF, GGA$).

Finally, a lot of effort is being directed towards the development of dispersion-corrected XC functionals accounting for the non-local dispersion (van der Waals) interactions. Examples of these range from empirical or environment-dependent $1/r^6$ (with r the distance) terms up to long-range density functionals and more sophisticated 'beyond pairwise additivity' schemes [42–46]. Enlightening reviews on these approaches have been recently published [47, 48]. Despite the complexity of an XC functional it is not always efficient. Higher order functionals could be less efficient than LDA for simple systems. Note, that in approximate XC functionals the cancellation of the self-interaction term is incomplete leading to a self-interaction (SI) error. It is mandatory to first test the XC functionals before applying them to the target systems.

2.4.2 Pseudopotentials

A simplification of the complexity of real systems within the DFT simulations is possible through pseudopotentials (PS). These account for the complex behavior of the all-electron case by involving only the valence electrons. The single-particle equations for the all-electron picture are replaced by equations which involve only the valence electrons:

$$\left[-\frac{\hbar^2}{2m_e}\nabla^2 + V_{eff}^{pseudo}\right]\psi_i^{pseudo} = \varepsilon_i^{pseudo}\psi_i^{pseudo} \quad (2.24)$$

In this equation, the effective potential (V_{eff}) and the single-particle wavefunctions (ψ_i) are replaced by the pseudopotential (V_{eff}^{pseudo}) and the single-particle pseudo-wavefunctions (ψ_i^{pseudo}) (see figure 2.3). The core states are eliminated and the valence electrons are described by pseudo-wavefunctions with fewer nodes making their computation easier. The core electrons and the nuclei are considered as rigid non-polarizable cores and only the valence electrons are chemically active. An important aspect in defining a pseudopotential is the cutoff radius, r_c in figure 2.3.

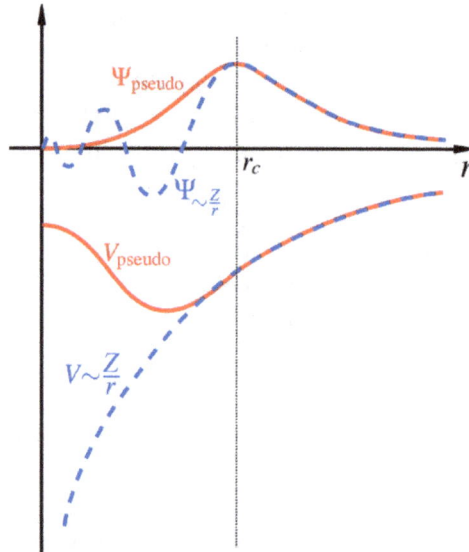

Figure 2.3. The concept of pseudopotentials is sketched. The pseudopotential (V_{pseudo}) and pseudo-wave-function (Ψ_{pseudo}) are compared to the real Coulomb potential of the nucleus and the respective all-electron wavefunction. Above the cutoff radius r_c, both the potentials and wavefunctions should match (source: https://en.wikipedia.org/wiki/Pseudopotential).

Above this distance from the center of the atom, the pseudo- and all-electron valence eigenstates are the same. A larger cutoff makes the pseudopotential softer (more rapidly converging). Soft pseudopotentials, though, are typically less transferable and behave differently in various systems.

Similar to the work done on XC functionals, a lot of different approaches for approximating pseudopotentials are developed. Representative types of pseudopotentials involve the orthogonalized plane waves (OPW) (using a smooth part of valence function with core-like functions) [49], the norm-conserving pseudopotentials (in which norm-conserving pseudo functions are normalized and are solutions of a model potential chosen to reproduce valence properties of an all-electron calculation) [50, 51], the l-dependent norm-conserving pseudopotentials (treating independently each l, m state) [52], the ultra soft pseudopotentials (using a smooth and an auxiliary function around each ion core that represents the rapidly varying part of density) [53], and projector augmented waves (PAW) (transforming rapidly oscillating functions near the core into smooth wavefunctions) [54, 55]. Nonlinear core [56] or semicore electron corrections [57] can take into account the overlap between the core and valence wavefunctions. Issues related to PS are the unscreening, the core corrections, the transferability, and their hardness. The choice of the pseudopotential is closely related to the XC functional taken and is not always equally efficient for all systems involving the same types of atoms, and optimization of these is being attempted [58].

2.4.3 Basis sets

A further important aspect in quantum mechanical simulations is the choice of basis sets. Typically, a finite set of basis functions—a collection of vectors defining the phase space—is used. The wavefunctions of a system are vectors. The components of these vectors correspond to coefficients in a linear combination of the basis functions in the chosen basis set. In principle, a basis set is the set of one-particle wavefunctions used to build molecular orbitals. For the calculations, atomic-like orbitals or plane waves can be used as basis sets. For the former, the wavefunctions are centered around an atom. For the latter, the wavefunctions are centered around a grid space in the reciprocal space. It is also possible to use plane waves centered around a grid point in the real space. In principle, the basis set is incomplete. The challenge is to use basis sets which would lead to a correct description of the physical system despite their incompleteness.

Here, the discussion on the basis-sets will be restricted to their main aspects. The basis sets differ in their size and complexity. The smaller the basis set and less complex, the faster the calculations typically are. Often, in these cases the reliability of the results is rather low. Throughout the years, different flavors of basis sets have been proposed [59]. Atomic-like orbitals can be distinguished as Slater-type orbitals (STOs), Gaussian-type orbitals (GTOs) or contracted Gaussian-type orbitals (CGTOs). The basis sets can be minimal (describing only the most basic aspects of the orbitals) or extended (including a more detailed description). The minimal basis sets include one basis functions (STOs, GTOs, or CGTOs) for each atomic orbital in the atom. The extended basis sets can be divided into double-zeta (DZ), triple-zeta (TZ), quadruple-zeta (QZ), split-valence (SV), polarized (P) or diffuse (D) sets. In DZ basis sets, each atomic orbital is expressed as the sum of two STOs. This is a much more accurate description as in reality the orbitals do not have the same shape as supposed in the minimal basis sets. With the DZ sets, each orbital is treated separately. Similarly, the TZ basis sets treat the atomic orbitals as the sum of three STOs, the QZ basis, as the sum of four STOs, and so on.

Note, that basis sets in the form '6-31++G', where the digits denote the exact features of the set, are typically used in quantum chemistry. In physics, the basis sets commonly used are the DZP, TZP, etc according to the notation above. Due to the use of a finite set of atomic orbitals and the fact that atoms in molecules interact (hence their basis orbitals overlap) the calculations are prone to the basis set superposition error (BSSE). These errors can be overcome using two methods: the chemical Hamiltonian approach (CHA) [60] and the counterpoise method (CP) [61]. In CHA, the mixing of the basis sets is prevented by using a Hamiltonian in which all the elements including projector terms have been removed. CP corrects for the BSSE *a posteriori*. All calculations are repeated using ghost orbitals[3] and the errors are subtracted from the uncorrected energy.

---

[3] Ghost orbitals are basis functions without electrons or protons.

2.4.4 Quantum transport calculations (DFT + non-equilibrium Green's functions)

As discussed, DFT can efficiently describe ground state electronic properties of physical systems. When combined with another method, known as non-equilibrium Green's functions (NEGFs) it can also evaluate the electronic transport properties across structures [62, 63]. From the theory side, this involves the description of coupling two metallic parts (the leads/electrodes) and the introduction of electronic states which can accommodate the flow or hopping of electrons. This method is the basic quantum-mechanical method used to investigate steady-state transport, electronic transmission, and the current–voltage (*IV*) characteristics (at low voltages) of small structures, electronic devices, and molecular devices where molecules are placed between two metallic leads [63].

The concept of the DFT–NEGF scheme is to divide the system into three different regions, the left (L), right (R) semi-infinite electrodes and the central (scatter) region. The Kohn–Sham Green's functions G_L and G_R represent the L and R leads. The scatter region is described by a Kohn–Sham equation in which the coupling to the leads is also taken into account. This is done through the non-Hermitian self-energy operators:

$$\Sigma_L = H_{CL} G_L H_{LC} \tag{2.25a}$$

$$\Sigma_R = H_{CR} G_R H_{RC} \tag{2.25b}$$

where H_{CL}, H_{LC}, H_{CR}, H_{RC} are the blocks of the Kohn–Sham Hamiltonian matrix, which connect the scatter region with the L and R electrodes. Knowledge of the self-energy operators can lead to the transmission function $\mathcal{T}(E)$ of the system with respect to the energy E:

$$\mathcal{T}(E) = tr\left[\Gamma_L G(E) \Gamma_R G(E)^\dagger\right] \tag{2.26}$$

where $\Gamma_j = i(\Sigma_i - \Sigma_i^\dagger)$ with $j = L, R$ are the non-Hermitian parts of the self-energies Σ_j and G the Green's functions of the total Kohn–Sham system. The transmission function with the aid of the Landauer formula connects to the *IV* characteristics of the system at a voltage V:

$$I(V) = \frac{e}{\pi\hbar} \int \mathcal{T}(E, V)\left[f(E, \mu_L) - f(E, \mu_R)\right] dE \tag{2.27}$$

$f(E, \mu_L)$ and $f(E, \mu_R)$ are the local Fermi–Dirac distributions of the two leads based on the chemical potentials μ_L and μ_R that relate to the L and the R lead, respectively. Details on the DFT–NEGF method and its applications can be found elsewhere [64–66]. The electronic transport cannot always be considered a steady state and needs to be treated with care (e.g. using XC potentials with a derivative discontinuity) [67].

2.5 Time-dependent density functional theory (TDDFT)

DFT is a ground state method in which the Kohn–Sham equations lead to independent particle eigenvalues, which do not correspond to true electron removal or addition energies. Accordingly, eigenvalue differences do not correspond to excitation energies. The time-dependent DFT (TDDFT) extends DFT to the treatment of excitations or more general time-dependent phenomena [68–70]. Since in the full many-body problem excitations are described in terms of response functions, a theory needs to be constructed which is able to describe the dynamical density response within the Kohn–Sham framework. The electrons feel the time-dependent Kohn–Sham effective potential. The TDDFT scheme is quite general, but includes two regimes in which the time dependent potential is: (a) weak or (b) strong. In (a) one should resort to linear-response theory and calculate for example the optical absorption spectra. In (b) a full solution of the Kohn–Sham equations is necessary. This would be relevant to the treatment of atoms or molecules in strong laser fields. TDDFT can describe non-linear phenomena, such as multi-phonon ionization or high-harmonic generation.

A core ingredient of the TDDFT methods is the **Runge–Gross theorem** [68], which is the time-dependent extension of the Hohenberg–Kohn approach. The theorem states that if two systems with N electrons have the same initial state, but are subject to two different time-dependent potentials, their respective time-dependent electron densities will differ, meaning that the electron density is uniquely defined by the external potential. For a non-interacting system[4], the N single-particle orbitals obey the time-dependent Schrödinger equation:

$$i\hbar\frac{\partial}{\partial t}\phi_i(\vec{r}, t) = \left[-\frac{\hbar^2}{2m}\nabla + V_{KS}(\vec{r}, t)\right]\phi_i(\vec{r}, t) \tag{2.28}$$

with $\phi_i(\vec{r}, 0) = \phi_i(\vec{r})$. The external local potential V_{KS} is chosen such that the density of the Kohn–Sham electrons is the same as the density of the original interacting system. The electron density is defined as $n_s(\vec{r}, t) = \sum_{i=1}^{N}|\phi_i(\vec{r}, t)|^2$. The Runge–Gross theorem asserts that all observables can be calculated with the knowledge of the one-body density.

Since in time-dependent systems the energy is not a conserved property and based on the Dirac action, Runge and Gross defined a quantum mechanical action $A[\phi]$. This action depends on the single-particle wavefunctions ϕ. Given the unique mapping between densities and wave-function, Runge and Gross then treated the Dirac action as a density functional,

$$A[n] = A\left[\Psi[n]\right] \tag{2.29}$$

A formal expression for the exchange-correlation part of the action can be derived and the potential can be written as:

$$V_{KS}(\vec{r}, t) = V_{ext}(\vec{r}, t) + V_{Hartree}(\vec{r}, t) + V_{XC}(\vec{r}, t) \tag{2.30a}$$

---

[4] Recall, that according to the KS theory a non-interacting system with an interaction potential which is zero is taken to form an electron density which is equal to that of the interacting system.

with

$$V_{\text{Hartree}}(\vec{r}, t) = \int \frac{n(\vec{r}, t)}{|\vec{r} - \vec{r}'|} d\vec{r} \qquad (2.30b)$$

Using the Keldysh formalism a new action functional A_{XC} can be defined [71]:

$$V_{\text{XC}}(\vec{r}, t) = \frac{\delta A_{\text{XC}}}{\delta n(\vec{r}, \tau)} \Bigg|_{n(\vec{r},t)} \qquad (2.31)$$

with τ the Keldysh pseudo-time. In the exact theory, $V_{\text{XC}}[n](\vec{r}, t)$ is a functional of $n(\vec{r}', t')$ for all earlier times $t' \leqslant t$, which needs functionals that include non-local effects in time.

An alternative is to used the linear-response TDDFT, which does not need the full solution of the time-dependent Kohn–Sham equations and can be used when a small, time-dependent perturbation is applied to the system. The advantage of this theory is that to a first order approximation the common features of DFT can be used as the variation of the system depends only on the ground-state wavefunction. The methodology involves the expression of the linear response of the electron density and the expansion of the variation of the Hartree and exchange correlation functionals with respect to the variation in the electron density. Combining these expressions with the response equation of the Kohn–Sham system leads to the derivation of the excitation energies [69, 72–76]. The calculation of optical absorption spectra can be done with the linear-response TDDFT. The excitation energies within TDDFT are obtained through differences between the ground-state Kohn–Sham eigenvalues (by taking expectation values of the Hamiltonian for different states).

XC functionals in TDDFT. The only fundamental approximation in TDDFT is related to the description of the XC functionals. Contrary to DFT, the development of XC functionals for TDDFT is still at an early stage [68, 77]. The main XC functionals for TDDFT are reviewed next.

Adiabatic local density approximation [78]. The adiabatic local density approximation (ALDA) is reminiscent of LDA and can be expressed through the equation:

$$F_{\text{ALDA}}^{\text{XC}}(\vec{r}, t) = \Gamma_{\text{hom}}^{\text{XC}}\big(n(\vec{r})\big)\Big|_{n=n(\vec{r},t)} \qquad (2.32)$$

ALDA assumes that the functional at $\vec{r}$ and time t is equal to the XC functional of the static homogeneous electron gas with a density $n(\vec{r})$. In principle, it is the same functional as in the local density approximation, but is employed at each time with an electron density $n(\vec{r}, t)$. The XC functional depends only on the density at the same time and is local in time. ALDA is a simple description and retains all the problems of LDA. It works well for systems near resonance, where the functional should include the particular states involved in the transition, but fails in calculating ionization potentials.

Time-dependent optimized effective potential [79]. This functional is an alternative to orbital-dependent XC functionals. A typical time-dependent OEP is the exact exchange (EXX). EXX is still local, but is obtained through the solution of the extremely non-local and non-linear (Fock) integral equation, which is the action functional:

$$A_{\text{EXX}}^X = -\frac{1}{2} \sum_{j,k}^{\text{occ}} \int_{t_0}^{t_1} dt \int d^3r \int d^3r' \frac{\psi_j^*(\vec{r}', t)\psi_k(\vec{r}', t)\psi_j(\vec{r}, t)\psi_k^*(\vec{r}, t)}{|\vec{r} - \vec{r}'|} \tag{2.33}$$

The XC functional is then the functional derivative of the XC action with respect to the Kohn–Sham wavefunctions:

$$F_{\text{XC}}(\vec{r}, t) = \frac{1}{\psi_j^*(\vec{r}, t)} \frac{\delta A^{\text{XC}}[\psi_j]}{\delta \psi_j}$$

The functional is local, though derived through the non-local and non-linear integral equation. Another description for a time-dependent OEP functional is the one proposed by Krieger, Li and Iiofrate (KLI) [80], which simplifies the above derivation using a semi-analytic expression and is a very good approximation to EXX. Both the EXX and KLI functionals have the correct $1/r$ asymptotic behavior for neutral finite systems contrary to ALDA.

Functional with a memory [81]. Non-local XC functionals including memory effects from previous times have been proposed. An expression for such a functional has the form

$$F_{\text{XC}}(\vec{r}, t) = \frac{1}{n(\vec{r}, t)} \nabla \int dt' \, \Pi^{\text{XC}}\left(n\left(\vec{R}, t'\right), t - t'\right) \tag{2.34}$$

where Π^{XC} is a pressure-like scalar memory function of two variables determined by requiring it to reproduce the scalar linear response of a homogeneous electron gas. $\vec{R}$ is a trajectory with the boundary condition $\vec{R}(t|\vec{r}, t) = \vec{r}$ and $\vec{r}$ is an actual location. This XC functional is non-local in time, but still local in $\vec{r}$. Other XC functionals for XC are also being developed and try to include non-locality both in time and space [82].

2.5.1 Computational scaling

In the following, some computational aspects, such as the scaling of the system size N with the computational time, are reviewed for quantum-mechanical methods presented in this chapter. Note that N can be either the number of electrons or orbitals modeled with these methods. Only methods presented above are reviewed. N denotes the system size, which is typically the number of orbitals or electrons in the system.

- *Hartree–Fock (HF, SCF).* Within HF, only one many-electron Slater determinant is used and the calculations are *ab initio* within a mean field approximation. The scaling of the method is $\mathcal{O}(N^3 - N^4)$ and the error in the energy ≈ 15 kcal mol^{-1}.

- *Density functional theory (B3LYP, BLYP, PW91, PBE, ...).* DFT is a slightly empirical scheme due to the approximations used for the XC functionals. The error is ~4 kcal mol^{-1} and the scaling $\mathcal{O}(N^3)$. However, DFT implementations can reach a scaling of $\mathcal{O}(N)$ [83]. DFT is preferred for geometry optimizations, for the calculation of second derivatives, and transition-metal-containing systems.

- *Møller–Plesset (MP2, MP4, ...).* The scaling in these methods depends on the level of perturbation. MP2 is the cheapest MP method scaling as N^5 with the system size N. MP3 scales with N^6, while MP4, MP5, MP6, and MP7 scale with N^7, N^8, N^9, and N^{10}, respectively. The errors in MP2 are of the order of 5 kcal mol^{-1}.

- *Configurational interaction (CI, CISD, QCISD, ...).* The scaling depends on the number of orbitals and excitation levels considered. CISD and QCISD show a scaling of $\mathcal{O}(N^6)$ and $\mathcal{O}(N^{10})$, respectively. CISD, CISDT, and CISDTQ scale like $\mathcal{O}(N^6)$, $\mathcal{O}(N^8)$, and $\mathcal{O}(N^{10})$ with the system size N, respectively.

- *Coupled cluster (CCSD, CCSD(T), ...).* These have size-consistent solutions (two molecules calculated simultaneously have the same energy as two molecules calculated separately) due to the methodology on which they depend. CC is also very expensive as the scaling behaves like N^6 for CCSD, N^7 for CCSD(T), etc.

2.6 *Ab initio* MD and electronic structure

Up to this point, all methods presented are static ground state methods. The term 'static' refers to the fact that the BO approximation was used [84]: the motion of electrons and nuclei can be separated. This is a very good approximation, as long as the motion of nuclei is slow and the temperature is low. If this is not the case, the motion of the nuclei must be added by additionally performing molecular dynamics (MD) simulations using forces calculated from quantum electronic structure schemes. MD is based on solving Newton's equations of motion for classical particles. Here, MD will be presented only in view of modeling classically the dynamics of nuclei within QM schemes. The method will be reviewed in detail in chapter 3.

2.6.1 Calculation of forces in electronic structure simulations

A straightforward way to calculate the forces F_I between atoms, based on the ground state electronic wavefunctions ψ_0 would be through $\vec{F}_I = -\nabla_R \langle \psi_0 | H | \psi_0 \rangle$. Such a calculation would be too expensive and probably also inaccurate for dynamical simulations. Alternative schemes have been proposed for the calculation of the forces among the nuclei based on QM arguments.

Born-Oppenheimer molecular dynamics (BO MD) [84]. The Schrödinger equation for the ionic motion reads

$$\left[T_n + \varepsilon_n(\vec{R}) \right] \phi_n^n(\vec{R}) = E \phi_n^n(\vec{R}) \tag{2.35}$$

where the wavefunctions $\phi_n^n(\vec{R})$ correspond to the nuclei at positions $\vec{R}$ when electrons are in state $\psi_n^e(\vec{r})$ with $\varepsilon_n(\vec{R})$ the total electronic energy including an effective interaction $(\hat{V}_{NN})$ among the nuclei. The nuclei are viewed as point charges and the Schrödinger equation for the nucleic dynamics (2.35) can be replaced by:

$$\vec{F}_I = M_I \ddot{\vec{R}}_I = -\frac{\partial \varepsilon_n(\vec{R})}{\partial \vec{R}_I} = -\nabla_R \varepsilon_n(\vec{R})$$ (2.36)

calculated from the total electronic potential. M_I, $\vec{R}_I$ are the mass and position of the nuclei. On the DFT level this can be rewritten as:

$$-\nabla_R \varepsilon_n(\vec{R}) = -\nabla_R \underbrace{\sum_{J \neq I} \frac{Z_I Z_J}{|\vec{R}_I - \vec{R}_J|}}_{(1)} - \underbrace{\int n(\vec{r}) \nabla_R v_I(|\vec{r} - \vec{R}_I|) d\vec{r}}_{(2)}$$ (2.37)

where (1) is the Coulomb repulsion between nuclei, (2) is the Coulomb attraction exerted on nuclei by the electron cloud, $v_I(\vec{r})$ is the pseudopotential or $-\frac{Z_I}{r}$ in the all-electron approach, $n(\vec{r})$ is the electronic density of the ground state and r the position of the electrons.

Within a BO MD approach, each time step would solve for the static electronic structure given fixed nuclear positions at each time. That means, that the time-independent Schrödinger equation with classically propagating nuclei needs to be solved. The time-dependent electronic structure is a consequence of the nuclear motion:

$$M_I \ddot{\vec{R}}_I(t) = -\nabla_R \min_{\psi_0} \left\{ \langle \psi_0 \mid H_{el} \mid \psi_0 \rangle \right\}$$ (2.38)

which satisfies the Schrödinger equation for the electronic ground state $E_0 \psi_0 = H_{el} \psi_0$. In each MD step, the minimum of $\langle H_{el} \rangle$, that is, the electronic ground state, is reached.

Hellmann–Feynman forces. Without going into the details which can be found elsewhere [85, 86], the Hellmann–Feynman theorem states that

$$\frac{dE_\lambda}{d\lambda} = \int \psi^*(\lambda) \frac{dH_\lambda}{d\lambda} \psi(\lambda) d\vec{r}$$ (2.39)

where λ is a continuous parameter on which the Hamiltonian H_λ, its wavefunctions $\psi(\lambda)$, and the eigenenergy E_λ depend. In this way, the derivative of the total energy with respect to λ is related to the expectation value of the derivative of the Hamiltonian, which is defined through λ. The forces can be calculated once the electron density and its distribution in space are known from solving Schrödinger equation. The parameter λ typically denotes the coordinates for the nuclei when used for the calculation of intramolecular forces. Since, the exact ground state is not known, the Hellman–Feynman are not the exact forces. The theorem can also be written for the time-dependent case.

Ehrenfest molecular dynamics. The Ehrenfest forces are calculated by solving a coupled set of nuclear and electronic propagation equations:

$$M_I \ddot{\vec{R}}_I(t) = -\nabla_R \langle \psi_0 \mid H_{el} \mid \psi_0 \rangle \tag{2.40a}$$

$$i\hbar \frac{\partial \psi}{\partial t} = \left[-\sum_i \frac{\hbar^2}{2m_e} \nabla_r^2 + \varepsilon\left(\{\vec{r}_i\}, \{\vec{R}_I\} \right) \right] \psi \tag{2.40b}$$

Equations (2.40a) and (2.40b) describe the real time-dependent evolution of both the electronic and nucleic degrees of freedom. The equations also include non-adiabatic effects like the transitions between different electronic states. Contrary to the BO MD, reviewed above, the minimum of $\langle H_{el} \rangle$ is not reached in each MD step. In Ehrenfest MD, a wavefunction that minimizes $\langle H_{el} \rangle$ will remain in the minimum as the nuclei move.

2.6.2 Car–Parrinello MD

A very important acceleration in performing *ab initio* MD is possible through the Car–Parinello Molecular Dynamics (CPMD) scheme [87]. Within CPMD, not only are nuclear positions calculated using MD algorithms, but also electronic states. The electronic structure does not relax exactly to the ground state of the actual configuration of nuclei, but the calculated electronic structure will follow closely the exact ground state. Recall that the total energy of a system can be written as a functional depending on the single-particle wavefunctions ψ_k and the nuclear coordinates (all collected in a variable S):

$$E_{tot} = E_{tot}\left(\{\psi_k\}, S \right) = E_{tot}\left(\{c_{rk}\}, S \right) \tag{2.41}$$

where a finite basis set $\{\xi_r\}$ is used to express ψ_k as $\psi_k(\vec{r}) = \sum_r c_{rk}\xi_k(\vec{r})$.

The CPMD method uses the expression in equation (2.41) together with the orthonormality constraint as a starting point for obtaining the equilibrium (minimal energy) conformation. The electronic structure does not have to be calculated exactly for each conformation of the nuclei, as both electronic orbitals and nuclear positions are varied simultaneously in order to find the minimum. This leads to a considerable acceleration of the computations and opens up the dynamics simulations of larger systems at the quantum level. For the time dependence of the nuclear coordinates and their fictitious time dependence on the electronic wavefunctions, a Langrangian based on ψ_k and S and their time derivatives can be written:

$$\mathcal{L}\left(\{\psi_k\}, S \right) = \frac{\mu}{2}\sum_k \dot{\psi}_k^2 + \sum_n M_n \frac{\dot{\vec{R}}^2}{2} - E_{tot}\left(\psi_k, S \right) + \sum_{k,l} \Lambda_{kl} \langle \psi_k | \psi_l \rangle \tag{2.42}$$

and the problem is nailed down to a classical mechanics problem with the energy acting as a potential. The addition of a friction term to the equations of motion of this classical system, will lead to the relaxation of the degrees of freedom to values corresponding to the minimum of the classical potential. This would correspond to the energy of a quantum system at the equilibrium configuration of the nuclei.

In equation (2.42), M_n is the actual mass of the nth nucleus at a position $\vec{R}_n$ and $\mu < M_n$ is a small mass, which controls how well electronic wavefunctions adapt to the changing nuclear configurations. Different choices of μ lead to different rates of convergence towards an energy minimum. For $\mu \rightarrow 0$, the Lagrangian is the true Lagrangian of the system. The Lagrangian can be used to find the minimum of the total energy, but also to perform real MD at a finite temperature. When the nuclei move, the method might not have produced the minimal energy for the electrons before the next nuclear displacement and the electronic structure will lag behind the nuclear motion. This retardation effect implied by the Car–Parrinello Lagrangian does not seem to be related to a real physical behavior.

The CPMD method can be implemented using the Euler–Lagrange equations:

$$\mu \ddot{\psi}_k = -\frac{\partial E_{\text{tot}}}{\partial \psi_k} + 2 \sum_l \Lambda_{kl} \psi_l(\vec{r}) \tag{2.43a}$$

$$M_n \ddot{\vec{R}}_n = -\frac{\partial E_{\text{tot}}}{\partial \vec{R}_n} + \underbrace{\sum_{k,l} \Lambda_{kl} \frac{\partial \langle \psi_k | \psi_l \rangle}{\partial \vec{R}_n}}_{\rightarrow 0 \text{ if basis functions do not depend on } S} \tag{2.43b}$$

When the simulation starts with nuclear configurations too far from the equilibrium, it could be possible to end up with a local minimum and not the global one, and additional schemes would be needed to hop over local barriers and reach the global minimum. Overall, CPMD is a first-principles MD approach, which similarly to ground-state electronic structure methods involves plane wave basis sets, pseudo-potentials, etc. Within CPMD, fictitious dynamics are assumed to keep electrons close to the ground state.

2.7 Semi-empirical methods

In this section, a common semi-empirical scheme for the simulation of physical systems is presented. The term 'semi-empirical' denotes that these schemes are not *ab initio* schemes, but also not completely classical. Semi-empirical methods are typically based on first-principles assumptions or the matrix representation from QM, but also use empirical parameters obtained from experimental or *ab initio* data. The values of the matrix elements are found through empirical formulas that estimate the degree of overlap of specific atomic orbitals. The matrix is then diagonalized to determine the occupancy of the different atomic orbitals and empirical formulae are used once again to determine the energy contributions of the orbitals. There exists a wide variety of semi-empirical potentials, known as tight-binding potentials, which vary according to the atoms and systems being modeled. Other semi-empirical methods, such as the modified neglect of diatomic overlap (MNDO), which adds two center integrals for the repulsion between a charge distribution on one center and a charge distribution on another center in the formalism [88–90], the Austin model 1 (AM1) [91], the parameterized model 3 (PM3) [92], etc do exist and attempt to better optimize the empirical parameters.

2.7.1 The tight-binding scheme

The tight-binding (TB) scheme is a semi-empirical method developed for the calculation of electronic band structures in solid-state physics. TB uses an approximate set of wavefunctions based upon the superposition of wavefunctions for isolated atoms located at each atomic site (close to the linear combination of atomic orbitals (LCAO) method). TB gives good quantitative results and can be combined with other methods, for example MD to add the dynamics of a system or with random phase approximation (RPA) to add the dynamic response of a system. The advantage of TB over first-principle methods is that the former can deal with a larger number of particles than the latter. TB is also more accurate than classical methods, but is still a parameterized scheme.

The term 'tight-binding' describes the properties of the tightly bound electrons in solids. The electrons are tightly bound to the atom to which they belong with limited interactions with states on surrounding atoms. In this respect, the wavefunction of an electron will be rather similar to the atomic orbital of the free atom it belongs to and the energy of the electron will be close to the ionization energy of the electron in the free atom or ion. TB defines the one-particle tight-binding Hamiltonian in which the inter-atomic matrix elements are very important. The main contribution to the TB comes from Slater and Koster [93] and is a basic approximation for strongly correlated materials.

The starting point of TB is the Schrödinger equation

$$H|\psi\rangle = \varepsilon|\psi\rangle \tag{2.44}$$

where H the Hamiltonian of the system and its wavefunction $|\psi\rangle$ is expressed through the basis functions. These are the atomic-like orbitals $\{\phi_a\}$ defined as

$$|\psi\rangle = \sum_{i,a}^{N_b} c_{ia}|\phi_a\rangle \tag{2.45}$$

In this equation, N_a is the number of atoms, N_0 is the number of basis functions/atom, and $N_b = N_0 \cdot N_a$ the dimension of the phase-space. a and i are indices running over atoms and orbitals, respectively. The weights of each orbital at each atomic site is $c_{ia} \in \mathbb{C}$.

The orbitals are centered around the atoms and decay rapidly. Accordingly, the matrix elements between different orbitals are non-zero only for on-site terms and between pairs of neighboring atoms (within a cut-off distance). The minimum basis set is composed from states of the outer shell electrons. For example, for carbon, the four orbitals $|\phi_s\rangle$, $|\phi_{p_x}\rangle$, $|\phi_{p_y}\rangle$, $|\phi_{p_z}\rangle$ would be considered. In general, the basis is not orthonormal, and for the cross-term matrix elements, the overlap parameters S_{ab}, $\langle\phi_a|\phi_b\rangle = S_{ab} \neq \delta_{ab}$ are finite. The coefficients c_{ia} are obtained through a variational calculation and minimization of

$$G[\psi] = \int \psi^* H \phi \, \mathrm{d}\vec{r} - \varepsilon \int \psi^* \psi \, \mathrm{d}\vec{r} \tag{2.46}$$

$$\Rightarrow \sum_{j\beta} \left(H_{iaj\beta} - \varepsilon_i S_{a\beta} \right) c_{j\beta} = 0 \tag{2.47}$$

where $H_{i\alpha j\beta} = \langle\psi|H|\psi\rangle c_{i\alpha}^* c_{j\beta}\langle\phi_\alpha|H|\phi_\beta\rangle$ the matrix elements of the Hamiltonian and ε_i the eigenenergies. In a matrix notation:

$$\mathbb{H} = \varepsilon\mathbb{S} \tag{2.48}$$

In the end, the TB method is reduced to a diagonalization of the Hamiltonian and the calculation of $c_{i\alpha}$, ε_i.

A simple TB model for the total energy in the case of semiconductors would be: $E_{\text{tot}} = E_{\text{band}} + E_{\text{rep}}$, where E_{band} is the sum of the one-electron eigenvalues of the occupied states. This term includes the contribution due to outer electrons. $E_{\text{rep}} = \frac{1}{2}\sum_{ij} v_R(r_{ij})$ is the sum of the short range repulsive pair potentials $v_R(r_{ij})$, which is usually an empirical term. Defining as ρ the density matrix and starting from the sum of the eigenvalues as discussed, one gets:

$$E_{\text{band}} = 2\sum_s n_s\varepsilon_s = 2\sum_s n_s\langle\psi_s|H|\psi_s\rangle = \cdots = 2\,\text{Tr}[\rho H] \tag{2.49}$$

In the equations, the sum is taken over all the occupied states s. In the empirical TB, the matrix elements for S, H are not computed, but analytical parameterized expressions as functions of atomic orbitals are used. Again, for simple elements like C or Si only s, p orbitals are taken and four orbitals at each atomic site are considered. Accordingly, a 4×4 matrix is needed, which includes 16 parameters that need to be fitted. The number of parameters increases when additional orbitals are taken. In this case, the transferability of the TB method can be questionable. However, the number of parameters needed can be reduced by accounting for symmetry rules.

An algorithm for applying the TB method to a physical system would flow as follows: first the atomic positions are given as input, as well as the empirical parameters for the matrix elements. The Hamiltonian is then calculated based on the parameters used and is diagonalized according to equation (2.48). The eigenvalues and eigenvectors obtained from the diagonalization can lead to the electronic ground state. In principle, TB can lead to similar properties as HF and DFT, but includes many more approximations and empirical parameters. Due to these assumptions, TB can be more efficient in modeling larger systems, where the application of HF or DFT becomes computationally very demanding. A typical scaling for TB approaches is $\mathcal{O}(N^2)$, but can become also less ($\mathcal{O}(N)$). Additional information on the TB method and different TB schemes can be found elsewhere [15, 94–104]. Note, that TB can also be combined with MD in order to perform dynamic simulations [105] and also DFT [106] to increase the accuracy of the method.

References

[1] Anisimov V I 2000 *Strong Coulomb Correlations in Electronic Structure Calculations* (Boca Raton, FL: CRC Press)
[2] Launay J-P and Verdaguer M 2013 *Electrons in Molecules: From Basic Principles to Molecular Electronics* (Oxford: Oxford University Press)

[3] Jensen F 1990 A remarkable large effect of spin contamination on calculated vibrational frequencies *Chem. Phys. Lett.* **169** 519–28

[4] Stanton J F 1994 On the extent of spin contamination in open-shell coupled-cluster wave functions *J. Chem. Phys.* **101** 371–4

[5] Andrews J S *et al* 1991 Spin contamination in single-determinant wavefunctions *Chem. Phys. Lett.* **183** 423–31

[6] Introduction to electronic structure methods by Rothlisberger U and Tavernelli I http://archive.org (accessed 1 August 2016)

[7] Hehre W J, Radom L, von P, Schleyer R and Pople J A 1986 *Ab Initio Molecular Theory* (New York: Wiley)

[8] Kaxiras E 2003 *Atomic and Electronic Structure of Solids* (Cambridge: Cambridge University Press)

[9] Martin R M 1999 *Electronic Structure: Basic Theory and Practical Methods* (Cambridge: Cambridge University Press)

[10] Hohenberg P and Kohn W 1964 Inhomogeneous electron gas *Phys. Rev.* **136** B864

[11] Kohn W and Sham L J 1965 Self-consistent equations including exchange and correlation effects *Phys. Rev.* **140** A1133

[12] Hartree D R *The Wave Mechanics of an Atom With a Non-Coulomb Central Field. Part I. Theory and Methods* vol 24 (Cambridge: Cambridge Universtiy Press)

[13] Fock V 1930 Näherungsmethode zur lösung des quantenmechanischen mehrkörperproblems *Z. Phys.* **61** 126–48

[14] Sharp R T and Horton G K 1953 A variational approach to the unipotential many-electron problem *Phys. Rev.* **90** 317

[15] Ashcroft N W and Mermin N D 1976 *Solid State Physics* (New York: Holt, Rinehart and Winston)

[16] Glaesemann K R and Schmidt M W 2010 On the ordering of orbital energies in high-spin ROHF *J. Phys. Chem.* A **114** 8772–7

[17] Amos T and Snyder L C 1964 Unrestricted HartreeFock calculations. I. an improved method of computing spin properties *J. Chem. Phys.* **41** 1773–83

[18] Purvis G D III and Bartlett R J 1982 A full coupled-cluster singles and doubles model: The inclusion of disconnected triples *J. Chem. Phys.* **76** 1910–8

[19] Raghavachari K, Trucks G W, Pople J A and Head-Gordon M 1989 A fifth-order perturbation comparison of electron correlation theories *Chem. Phys. Lett.* **157** 479–83

[20] Van Voorhis T and Head-Gordon M 2001 Two-body coupled cluster expansions *J. Chem. Phys.* **115** 5033–40

[21] Møller C and Plesset M S 1934 Note on an approximation treatment for many-electron systems *Phys. Rev.* **46** 618

[22] Schrödinger E 1926 Quantisierung als eigenwertproblem *Ann. Phys., Lpz.* **385** 437–90

[23] Rayleigh J W S 1944 *The Theory of Sound* vol 1 (New York: Dover)

[24] Callaway J 1991 *Quantum Theory of the Solid State* (London: Academic)

[25] Pople J A, Head-Gordon M and Raghavachari K 1987 Quadratic configuration interaction. a general technique for determining electron correlation energies *J. Chem. Phys.* **87** 5968–75

[26] Parr R G and Yang W 1989 Density-Functional Theory of Atoms and Molecules *International Series of Monographs on Chemistry* vol 16 (Oxford: Oxford University Press) pp 160–80

[27] Baker J, Scheiner A and Andzelm J 1993 Spin contamination in density functional theory *Chem. Phys. Lett.* **216** 380–8

[28] Langreth D C and Mehl M J 1983 Beyond the local-density approximation in calculations of ground-state electronic properties *Phys. Rev.* B **28** 1809

[29] Becke A D 1988 Density-functional exchange-energy approximation with correct asymptotic behavior *Phys. Rev.* A **38** 3098

[30] Perdew J P and Wang Y 1992 Pair-distribution function and its coupling-constant average for the spin-polarized electron gas *Phys. Rev.* B **46** 12947

[31] Perdew J P, Burke K and Ernzerhof M 1996 Generalized gradient approximation made simple *Phys. Rev. Lett.* **77** 3865

[32] Singh D J 1993 Weighted-density-approximation ground-state studies of solids *Phys. Rev.* B **48** 14099

[33] Perdew J P and Zunger A 1981 Self-interaction correction to density-functional approximations for many-electron systems *Phys. Rev.* B **23** 5048

[34] Anisimov V I, Aryasetiawan F and Lichtenstein A I 1997 First-principles calculations of the electronic structure and spectra of strongly correlated systems: the LDA+U method *J. Phys.: Condens. Matter* **9** 767

[35] Talman J D and Shadwick W F 1976 Optimized effective atomic central potential *Phys. Rev.* A **14** 36

[36] Kümmel S and Perdew J P 2003 Optimized effective potential made simple: orbital functionals, orbital shifts, and the exact Kohn–Sham exchange potential *Phys. Rev.* B **68** 035103

[37] Gu H- Q *et al* 2013 Slater approximation for Coulomb exchange effects in nuclear covariant density functional theory *Phys. Rev.* C **87** 041301

[38] Aryasetiawan F and Gunnarsson O 1998 The GW method *Rep. Prog. Phys.* **61** 237

[39] Lee C, Yang W and Parr R G 1988 Development of the Colle-Salvetti correlation-energy formula into a functional of the electron density *Phys. Rev.* B **37** 785

[40] Heyd J, Scuseria G E and Ernzerhof M 2003 Hybrid functionals based on a screened Coulomb potential *J. Chem. Phys.* **118** 8207–15

[41] Zhao Y and Truhlar D G 2006 Density functional for spectroscopy: no long-range self-interaction error, good performance for Rydberg and charge-transfer states, and better performance on average than B3LYP for ground states *J. Chem. Phys.* A **110** 13126–30

[42] Von Lilienfeld O A, Tavernelli I, Rothlisberger U and Sebastiani D 2004 Optimization of effective atom centered potentials for London dispersion forces in density functional theory *Phys. Rev. Lett.* **93** 153004

[43] Tran F and Hutter J 2013 Nonlocal van der Waals functionals: The case of rare-gas dimers and solids *J. Chem. Phys.* **138** 204103

[44] Grimme S 2006 Semiempirical GGA-type density functional constructed with a long-range dispersion correction *J. Comput. Chem.* **27** 1787–99

[45] Tkatchenko A and Scheffler M 2009 Accurate molecular van der Waals interactions from ground-state electron density and free-atom reference data *Phys. Rev. Lett.* **102** 073005

[46] Zimmerli U, Parrinello M and Koumoutsakos P 2004 Dispersion corrections to density functionals for water aromatic interactions *J. Chem. Phys.* **120** 2693–9

[47] Klimeš J and Michaelides A 2012 Perspective: advances and challenges in treating van der Waals dispersion forces in density functional theory *J. Chem. Phys.* **137** 120901

[48] Grimme S 2011 Density functional theory with London dispersion corrections *WIREs Comput. Mol. Sci.* **1** 211–28

[49] Callaway J 1955 Orthogonalized plane wave method *Phys. Rev.* **97** 933

[50] Hamann D R, Schlüter M and Chiang C 1979 Norm-conserving pseudopotentials *Phys. Rev. Lett.* **43** 1494

[51] Troullier N and Martins J L 1991 Efficient pseudopotentials for plane-wave calculations *Phys. Rev.* B **43** 1993

[52] Shirley E L, Allan D C, Martin R M and Joannopoulos J D 1989 Extended norm-conserving pseudopotentials *Phys. Rev.* B **40** 3652

[53] Vanderbilt D 1990 Soft self-consistent pseudopotentials in a generalized eigenvalue formalism *Phys. Rev.* B **41** 7892

[54] Blöchl P E 1994 Projector augmented-wave method *Phys. Rev.* B **50** 17953

[55] Kresse G and Joubert D 1999 From ultrasoft pseudopotentials to the projector augmented-wave method *Phys. Rev.* B **59** 1758

[56] Louie S G, Froyen S and Cohen M L 1982 Nonlinear ionic pseudopotentials in spin-density-functional calculations *Phys. Rev.* B **26** 1738

[57] Reis C L, Pacheco J M and Martins J L 2003 First-principles norm-conserving pseudo-potential with explicit incorporation of semicore states *Phys. Rev.* B **68** 155111

[58] Rappe A M, Rabe K M, Kaxiras E and Joannopoulos J D 1990 Optimized pseudopotentials *Phys. Rev.* B **41** 1227

[59] Basis sets. https://bse.pnl.gov/bse/portal (accessed 7 March 2016)

[60] Mayer I and Valiron P 1998 Second order Møller-Plesset perturbation theory without basis set superposition error *J. Chem. Phys.* **109** 3360–73

[61] Van Duijneveldt F B, van Duijneveldt-van de Rijdt J G C M and van Lenthe J H 1994 State of the art in counterpoise theory *Chem. Rev.* **94** 1873–85

[62] Datta S 1997 *Electronic Transport in Mesoscopic Systems* (Cambridge: Cambridge University Press)

[63] Datta S 2005 *Quantum Transport: Atom to Transistor* (Cambridge: Cambridge University Press)

[64] Cuniberti G, Fagas G and Richter K 2006 *Introducing Molecular Electronics: A Brief Overview* (Berlin: Springer)

[65] Di Ventra M 2008 *Electrical Transport in Nanoscale Systems* vol 14 (Cambridge: Cambridge University Press)

[66] Rocha A R and Sanvito S 2010 *Electronic Transport at the Nanoscale: Theoretical and Computational Aspects* (Dublin: LAP Lambert Academic Publishing)

[67] Kurth S, Stefanucci G, Khosravi E, Verdozzi C and Gross E K U 2010 Dynamical Coulomb blockade and the derivative discontinuity of time-dependent density functional theory *Phys. Rev. Lett.* **104** 236801

[68] Runge E and Gross E K U 1984 Density-functional theory for time-dependent systems *Phys. Rev. Lett.* **52** 997

[69] Marques M A L, Maitra N T, Nogueira F M S, Gross E K U and Rubio A 2012 *Fundamentals of Time-Dependent Density Functional Theory* vol 837 (Berlin: Springer)

[70] Ullrich C A and Yang Z- H 2014 A brief compendium of time-dependent density functional theory *Braz. J. Phys.* **44** 154–88

[71] Vignale G 2008 Real-time resolution of the causality paradox of time-dependent density-functional theory *Phys. Rev.* A **77** 062511

[72] Ullrich C A 2011 *Time-Dependent Density-Functional Theory: Concepts and Applications* (Oxford: Oxford University Press)

[73] Maitra N T, Zhang F, Cave R J and Burke K 2004 Double excitations within time-dependent density functional theory linear response *J. Chem. Phys.* **120** 5932–7

[74] Petersilka M, Gossmann U J and Gross E K U 1998 Time-dependent optimized effective potential in the linear response regime *Electronic Density Functional Theory* (Berlin: Springer) pp 177–97

[75] Yabana K, Nakatsukasa T, Iwata J-I and Bertsch G F 2006 Real-time, real-space implementation of the linear response time-dependent density-functional theory *Phys. Status Solidi b* **243** 1121–38

[76] Casida M E 2005 Propagator corrections to adiabatic time-dependent density-functional theory linear response theory *J. Chem. Phys.* **122** 054111

[77] Marques M A L and Gross E K U 2004 Time-dependent density functional theory *Annu. Rev. Phys. Chem.* **55** 427–55

[78] Helbig N *et al* 2011 Density functional theory beyond the linear regime: Validating an adiabatic local density approximation *Phys. Rev.* A **83** 032503

[79] Ullrich C A, Gossmann U J and Gross E K U 1995 Time-dependent optimized effective potential *Phys. Rev. Lett.* **74** 872

[80] Krieger J B, Li Y and Iafrate G J 1992 Construction and application of an accurate local spin-polarized Kohn–Sham potential with integer discontinuity: exchange-only theory *Phys. Rev.* A **45** 101

[81] Dobson J F, Bünner M J and Gross E K U 1997 Time-dependent density functional theory beyond linear response: An exchange-correlation potential with memory *Phys. Rev. Lett.* **79** 1905

[82] Vignale G, Ullrich C A and Conti S 1997 Time-dependent density functional theory beyond the adiabatic local density approximation *Phys. Rev. Lett.* **79** 4878

[83] Soler J M *et al* 2002 The siesta method for ab-initio order-N materials simulation *J. Phys.: Condens. Matter* **14** 2745

[84] Born M and Oppenheimer R 1927 Zur quantentheorie der molekeln *Ann. Phys., Lpz.* **389** 457–84

[85] Hellman H 1939 Einführung in die quantenchemie (Deuticke, Leipzig, 1937); R P Feynman *Phys. Rev.* **56** 340

[86] Feynman R P 1939 Forces in molecules *Phys. Rev.* **56** 340

[87] Car R and Parrinello M 1985 Unified approach for molecular dynamics and density-functional theory *Phys. Rev. Lett.* **55** 2471

[88] Dewar M J S and Storch D M 1985 Development and use of quantum molecular models. 75. comparative tests of theoretical procedures for studying chemical reactions *J. Am. Chem. Soc.* **107** 3898–902

[89] Dewar M J S and Thiel W 1977 Ground states of molecules. 38. the mndo method. approximations and parameters *J. Am. Chem. Soc.* **99** 4899–907

[90] Dewar M J S *et al* 1985 Development and use of quantum mechanical molecular models. 76. AM1: a new general purpose quantum mechanical molecular model *J. Am. Chem. Soc.* **107** 3902–9

[91] Hestenes M R and Stiefel E 1952 Methods of conjugate gradients for solving linear systems *J. Res. Natl. Bur. Stand.* **49** 409–36

[92] Stewart J J P 1989 Optimization of parameters for semiempirical methods I. method *J. Comput. Chem.* **10** 209–20

[93] Slater J C and Koster G F 1954 Simplified LCAO method for the periodic potential problem *Phys. Rev.* **94** 1498

[94] Harrison W A 1980 *Electronic Structure and The Properties of Solids: The Physics of The Chemical Bond* (New York: Dover)

[95] Goringe C M *et al* 1997 Tight-binding modelling of materials *Rep. Prog. Phys.* **60** 1447

[96] Altland A and Simons B 2006 Interaction effects in the tight-binding system *Condensed Matter Field Theory* (Cambridge: Cambridge University Press)

[97] Harrison W A 1994 Tight-binding methods *Surf. Sci.* **299** 298–310

[98] Andriotis A N 1995 The scaling of the tight-binding Hamiltonian *J. Phys.: Condens. Matter* **7** L61

[99] Schulz S and Czycholl G 2005 Tight-binding model for semiconductor nanostructures *Phys. Rev.* B **72** 165317

[100] Lee Y-S, Nardelli M B and Marzari N 2005 Band structure and quantum conductance of nanostructures from maximally localized wannier functions: the case of functionalized carbon nanotubes *Phys. Rev. Lett.* **95** 076804

[101] Yamada H and Iguchi K 2010 Some effective tight-binding models for electrons in DNA conduction: a review *Adv. Cond. Matter Phys.* **2010** 380710

[102] Cohen R E, Mehl M J and Papaconstantopoulos D A 1994 Tight-binding total-energy method for transition and noble metals *Phys. Rev.* B **50** 14694

[103] Papaconstantopoulos D A and Mehl M J 2003 The Slater–Koster tight-binding method: a computationally efficient and accurate approach *J. Phys.: Condens. Matter* **15** R413

[104] Tang M S, Wang C Z, Chan C T and Ho K M 1996 Environment-dependent tight-binding potential model *Phys. Rev.* B **53** 979

[105] Wang C Z and Ho K M 1996 Tight-binding molecular dynamics for materials simulations *Journal of Computer-Aided Materials Design* **3** 139–48

[106] Kuwahara T *et al* 2016 Origin of Chemical Order in a-Si x C y H z: Density-functional tight-binding molecular dynamics and statistical thermodynamics calculations *J. Phys. Chem.* C **120** 2615–27

Chapter 3

Atomistic methods

Moving up in the scales in figure 1.2 allows modeling larger systems with a less accurate description than was described in chapter 2. Going away from the QM description moves further to a more classical description. A very common method at this level is known as Molecular Dynamics (MD).

3.1 Classical molecular dynamics

In principle, the experimental systems are too large to be determined by summing over all accessible states in a computer. Within MD (and Monte Carlo to be discussed in chapter 6), the physical quantities of a many-particle system are determined as statistical ensemble averages over a restricted set of states. In a quantum mechanical approach a full treatment of all interactions is computationally extremely expensive. QM is practical only for short time and small length scales, few particles and low temperature. MD takes over for longer time and length scales and can deal with finite temperatures. Overall, MD is a computational technique for computing equilibrium and dynamical properties of a **classical** many-body system. The theoretical basis for MD is analytical mechanics (Euler, Hamilton, Lagrange, Newton). The system moves in space along its physical trajectory as determined by Newton's equations of motion, which are integrated within MD and the simulated system develops over a period of time.

3.1.1 Basics of MD simulations

The MD method enables the connection of the microscopic structure to macroscopic properties. MD follows the dynamics (motion) of all atoms in the physical system and models N particles classically in a rectangular box of size $L_1 \times L_2 \times L_3$. The particles are assumed to interact with each other and Newton's equations of motion can be solved numerically to find these interactions:

$$\vec{F}_i = m_i \vec{a}_i \Rightarrow \begin{cases} \dot{\vec{r}}_i = \dfrac{\vec{p}_i}{m_i} \\ \dot{\vec{p}}_i = \vec{F}_i \end{cases} \tag{3.1}$$

where m_i, p_i, $\vec{a}_i$ are the mass, momentum, and acceleration of particle i. The forces acting on the particles explicitly relate to the inter-particle interactions modeled through the total potential energy $E\left(\{\vec{r}_j\}\right)$ calculated from inter-particle potentials:

$$\vec{F}_i = -\nabla E\left(\{\vec{r}_j\}\right) \tag{3.2}$$

In order to use MD, the initial positions of the particles and the inter-particle potentials need to be known to advance the system of particles in time and obtain its properties. A simple MD algorithm realizing this is sketched in figure 3.1. The MD simulations finish once the desired simulation time has been reached.

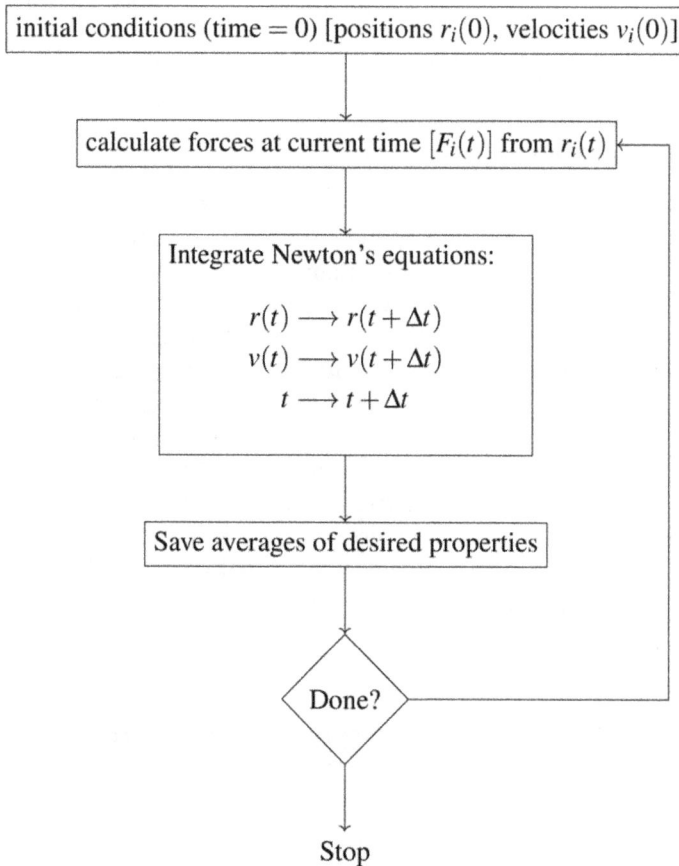

Figure 3.1. A simple minimal MD code.

Turning to Hamilton's picture, a Hamiltonian H can be defined for the system:

$$H\left(\{\vec{r}_i\}, \{\vec{p}_i\}\right) = V(\{\vec{r}_i\}(t)) + \sum_{i=1}^{3N} \frac{p_i^2(t)}{2m_i} \qquad (3.3)$$

where i denotes the particle identity, the frist term the potential and the second the kinetic energy. The Hamiltonian can lead to the derivation of the equations of motion:

$$\dot{\vec{r}}_i = \frac{\partial H}{\partial p_i}, \qquad \dot{\vec{p}}_i = -\frac{\partial H}{\partial r_i} \qquad (3.4)$$

and link the microscopic states to macroscopic properties. Given a thermodynamic state of the system, the method can evaluate the probabilities of finding a system in various microscopic states. In order to enable this link, statistical mechanics approaches and statistical ensembles are used. The micro-canonical, canonical, and isobaric/isothermal ensembles are the ones typically used. For example, in an MD simulation at constant volume V and temperature T (the NVT canonical ensemble), it is expected that the average kinetic energy corresponds to the desired temperature (recall the equipartition of the kinetic energy $\langle K \rangle = \frac{f}{2}k_BT$ for f degrees of freedom) and the dynamics is consistent with the canonical distribution. Another issue is the fact that all time averages are evaluated over a finite time. When a system is well equilibrated, this can be a very good approximation. Another important issue is the time step and the number of integration steps taken. When the correlation time of a system is smaller than the total simulation time, the numerical integration is not infinitely accurate. This hints to an optimum choice of computational speed and accuracy: the longer the integration time step, the more inaccurate the results. In such cases, the system would follow a trajectory in phase space deviating from that followed in reality). The MD method offers a direct simulation of a many-particle system, but the choice of the simulation parameters needs to be done with care.

3.1.2 Boundary conditions

In the following, focus will be given to implementation details of atomistic simulations. A first important issue is the fact that the system size in all simulations is smaller than in reality. This is not a big problem if the correlation length is much smaller than the system size. However, finite-size scaling should be considered when the correlation length exceeds the computational system size. In general, the finiteness of a system is defined through a boundary. The most common way to deal with this is to apply periodic boundary conditions (PBCs). In this way, the particles in the simulation box are virtually periodically repeated in space to form an infinite system. With PBCs the system of interest is surrounded by similar systems with exactly the same configuration of particles at any time and the interaction between two particles i, j is given by the equation

$$\vec{F}_{\mathrm{PBC}}(\vec{r}_j - \vec{r}_j) = \sum_n \vec{F}\left(\left| r_i - r_j + \sum_{\mu=1}^{3} L_\mu n_\mu \right|\right) \qquad (3.5)$$

where, L_μ are vectors along the edges of the volume of the rectangular system and the sum is taken over all vectors $\vec{n}$ with integer coefficients n_μ. The inter-particle force $\vec{F}$ is directed along the line connecting particle i and its image particle j. A sketch of the PBC scheme is shown in figure 3.2. The application of boundary conditions such as PBC is quite time consuming, as it involves the calculation of terms of an infinite sum until convergence is reached. This though can be easily resolved with neighbour lists and other schemes [1–3]. In such schemes, not all the particle images are taken into account in the interactions, but a list of the nearest neighbors for each particle in the simulations is made. Only the image particles in this list are then considered in calculating the forces in equation (3.5). Such approaches, considerably decrease the computational time.

3.1.3 Forces in molecular dynamics

The key point in MD simulations is the computation of the inter-particle forces, that is, the use of well optimized classical potentials. The interactions between particles are not known exactly and are approximated through classical pair potentials between particle centers. Often, these are named 'effective potentials' (a more

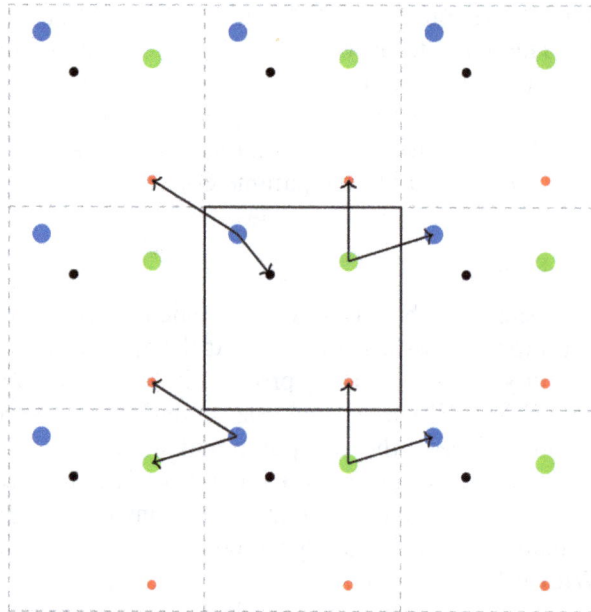

Figure 3.2. The simulation box and its particles are shown in the middle black rectangular box. PBCs are applied and repeat the central region in both directions in space. The arrows point to the image particles interacting with the particles in the simulation box.

extended discussion on potentials will be given in chapter 4). Typically, the inter-particle potentials $V(\vec{R})$ depend on the inter-particle distance $\vec{R}$ and are connected to the force acting on a particle $\vec{F} = -\dfrac{\mathrm{d}V(\vec{R})}{\mathrm{d}\vec{R}}$. The use of efficient potentials together with boundary conditions, leads to relatively fast computations, as the computational time for an N particle system shows a scaling of $\mathcal{O}(N^2)$. Proper optimization (using neighbor list, cut-offs etc.) can lead to $\mathcal{O}(N)$. The disadvantage of classical potentials or force fields is that these include empirical parameters and inherently carry errors. The potentials can be reliable only when they are applied to systems for which they have been developed. The quality of the potential parameters needs to be evaluated and often important factors such as the orbital interactions, polarizability, and multi-body effects are not represented through pair potentials.

3.2 Setting environment conditions

Most of the common quantum-mechanical simulation methods used in physics typically involve low temperatures, are done *in vacuo* and the influence of an environment cannot be easily accounted for. In classical dynamics simulations, such as MD, setting the environmental conditions can be explicitly done using thermo-stats (for applying a finite temperature) or barostats (for applying a pressure). If none of these is used, the MD simulations sample the microcanonical (NVE) ensemble, as this is the ensemble arising from the Newtonian dynamics which conserve the total energy.

3.2.1 Thermostats

The most common thermostats, ensuring that the desired temperature is applied to a system and the (NVT) ensemble is sampled, are reviewed here. A very informative and extended review on thermostats can be found elsewhere [4]. In principle, a thermostat is based on the modification of the Newtonian dynamics, in order to generate a statistical ensemble at a constant temperature and modulate the temper-ature of a system. It is then possible to enhance the efficiency in conformational search, avoid energy drifts caused by accumulation of numerical errors, and evacuate heat in dissipative non-equilibriium MD. The study of temperature-dependent processes (thermal coefficients, melting, etc) and matching experimental conditions can be efficiently done.

A variety of thermostats have been developed to add and remove energy from the boundaries of a system in a realistic way and approximating the canonical ensemble. The goal of a thermostat is not to keep the temperature constant (fix the kinetic energy), but to ensure that the average temperature is the desired one. At equilibrium (e.g. at constant temperature) the particle momenta (p) are distributed according to the Maxwell–Boltzmann distribution

$$P(p) = \left(\frac{\beta}{2Nm}\right)^{3/2} \mathrm{e}^{-\beta p^2/2m} \tag{3.6}$$

In cases at which the average kinetic energy per particle is kept fixed (in an isokinetic MD scheme or by velocity rescaling, discussed later) the simulated ensemble is not the correct (NVT).

The use of a thermostat requires the definition of an **instantaneous temperature** $\mathcal{T}$, which will be compared to a **reference temperature** T of a heat bath. For N_f degrees of freedom in a system, the average internal kinetic energy of the system relates to its macroscopic temperature (equipartition theorem) through $K = \langle \mathcal{K} \rangle = \frac{1}{2}k_B N_f T$ where $\mathcal{K}$ is the instantaneous kinetic energy and k_B the Boltzmann constant. At any time-step in the simulations, the instantaneous temperature is $\mathcal{T} = \frac{2}{k_B N_f}\mathcal{K}$. Maintaining an average constant temperature constant in MD simulations requires to impose control on the rate of change of these velocities through a modification of Newton's equations of motion $\ddot{r}_i(t) = \frac{F_i(t)}{m_i}$, with r_i, m_i the particle position and mass, and F_i the force acting on it. A common way to do this is through the Langevin equation, which serves as a prototype for many thermostats:

$$\ddot{r}_i(t) = \frac{F_i(t)}{m_i} - \underset{\underset{\text{atomic friction coefficient(positive)}}{\uparrow}}{\gamma(t)} \dot{r}_i(t) + m_i^{-1} \underset{\underset{\text{stochastic force}}{\uparrow}}{R_i(t)} \tag{3.7}$$

Note that the thermostats should recover the canonical ensemble and the relevant fluctuations in the temperature. Time-reversibility is also an important aspect of thermostats as the same particle trajectories should be taken if time could be reversed.

Langevin thermostat. In Langevin dynamics [2], the atoms in the system are assumed to be embedded in a sea of frictional particles of smaller size and the dynamics of the particles are described not by Newton's equation, but by the Langevin equation (3.7). The Langevin thermostat includes a viscous damping time, which is characteristic of the energy exchange between the particles of the system and a heat bath. This thermostat can get away with longer time steps than in an (NVE) ensemble and is a good choice for an initial equilibration of the system, but momentum transfer is destroyed.

Velocity (re)scaling. Rescaling the velocities of the particles is often an efficient way to obtain the desired temperature. Assuming again the equipartition theorem, it is easy to find a relation between the kinetic energy (or the particle velocities/positions) and the desired temperature. The simplest way to control the temperature is to multiply at each step the velocities with a factor $\lambda = \sqrt{\frac{T_0}{T(t)}}$, with $T(t)$ the actual and T_0 the desired temperature. Velocity rescaling is very simple and easily implemented, but does not follow the canonical ensemble, is not time-reversible and deterministic, and does not allow for the temperature fluctuations typically present in the canonical ensemble.

Berendsen thermostat. In order to allow temperature fluctuations, a weak coupling method to an external bath was introduced, and is known as the Berendsen thermostat [5]. This approach attempts to correct the deviations of the actual temperature $T(t)$ from the desired T_0 by multiplying the velocities (and momenta) by a certain factor in order to move the system dynamics towards the one corresponding to T_0. This factor is $\lambda = \sqrt{1 + \frac{\Delta t}{\tau_T}\left(\frac{T_0}{T} - 1\right)}$, where Δt is the MD time-step and τ_T is the time-coupling parameter to the heat bath. The motivation of the Berendsen thermostat is the minimization of local disturbances of a stochastic thermostat, while keeping the global effects the same. The starting point of the Berendsen thermostat is also the Langevin equation. This thermostat is deterministic and captures fluctuations in the temperature T that follow the ones expected for the (NVT) ensemble. Through this thermostat, a direct feedback to control the temperature can be achieved throughout the simulations. However, it does not generate a canonical ensemble and is not time-reversible.

Andersen thermostat. The Andersen thermostat [6] couples the system to a heat bath via the stochastic collisions of their particles. These collisions modify the kinetic energy of the particles. The collision-frequency ν describes the strength of the coupling. The probability of collision over time dt is νdt and the collisions are governed by the Poisson distribution $P(t) = \nu e^{-\nu t}$. Between collisions, the system evolves at constant energy. After a collision event, the new momentum of the particle is chosen from a Boltzmann distribution at temperature T. In the Andersen thermostat the equations of motion are Hamiltonian and include a stochastic collision term. This thermostat is the simplest NVT conserving thermostat and samples the canonical ensemble. The resulting trajectories are perfectly energy conserving.

Nosé–Hoover thermostat. The Nosé–Hoover thermostat [7, 8] is a deterministic method to keep the temperature around an average. The heat bath is an integral part of the system by adding an artificial variable associated with an artificial mass. The trajectory is deterministic and an extended Lagrangian formalism is used. The Hamiltonian of the system of N particles is given by:

$$H = \sum_{i=1}^{N} \frac{P_i^2}{2m_i s^2} + U(\vec{r}) + \frac{Q\dot{s}^2}{2} + (3N + 1)k_B T \ln s \tag{3.8}$$

where s is an additional degree of freedom, which marks the 'position' of an imaginary heat bath to which the system is coupled. s acts as an external system. p_s is the conjugate 'momentum' of this imaginary heat bath and Q the effective 'mass' associated with s, so that $p_s = Q\dot{s}$. Increasing Q increases the decay time of the response to instantaneous temperature jumps. These virtual variables relate back to the real particle variables. The Nosé–Hoover thermostat is easy to implement, it is deterministic, time reversible, and leads back to the canonical ensemble and the relevant fluctuations. However, it does not guarantee that the extended system is ergodic. Non-canonical fluctuations can arise as the system approaches the equilibrium if the thermostat is not properly and carefully implemented.

3.2.2 Barostats

In a manner similar to thermostats, barostats are used to perform simulations in an isothermal-isobaric (NPT) ensemble. In these simulations, the volume is a dynamic variable which changes during the simulation. The pressure of a classical N-body system can be calculated using the **Clausius virial theorem**:

$$P = \frac{2}{3V}(E_{\text{kin}} - \Xi) \tag{3.9}$$

where V is the box-volume, E_{kin} the kinetic energy in the system and Ξ the inner virial for pairwise additive interactions defined as $\Xi = \frac{1}{2}\sum_{i<j}\vec{r}_{ij} \cdot \vec{f}(r_{ij})$. In this equation, $\vec{f}(r_{ij})$ is the force between particles i and j separated by a distance r_{ij}. The pressure is actually not a scalar quantity but a second order tensor $\mathbb{P}$. Typically a correction of the pressure in the simulations can be achieved through a change in the inner virial Ξ by scaling the inter particle distances $\vec{r}_{ij}$. The pressure coupling occurs in a very similar way as with the temperature coupling. The existing barostats can be roughly divided to those assuming a constant box shape (but not volume) (e.g. Berendsen and Andersen) and those assuming a constant box volume (but not shape) (e.g. Parrinello–Rahman).

Other types of barostats than the ones reviewed here, have also been proposed. For example, the extended dimension of the Nosé–Hoover thermostat can be applied to a respective barostat. The Parrinello–Rahman barostat [9] extended this further. Within this barostat, each unit vector of the unit-cell is independent so that the volume is a variable in the simulation (similar to the Nosé–Hoover approach). The Parrinello–Rahman barostat also allows a dynamic box shape change, allowing control of the stress as well as the pressure. An extended presentation of available barostats can be found elsewhere [10].

Berendsen Barostat. Similar to the temperature coupling to an external bath, the Berendsen barostat adds an extra term to the equations of motion which leads to a pressure change of [5]:

$$\left(\frac{dp}{dt}\right)_{\text{bath}} = \frac{p_0 - p}{\tau_P} \tag{3.10}$$

where p_0, p are the target and actual (at the actual time step) pressure of the system. The parameter τ_P is a time constant controlling the coupling and thereby the pressure fluctuations. The larger this constant is, the weaker the coupling. The positions r and velocities v are rescaled through $\dot{r} = v + \alpha r$, where α is a parameter related to the isothermal compressibility and defines the scaling. The system volume V changes accordingly like $\dot{V} = 3\alpha V$. For an isotropic scaling, the same α values need to be used for the three x,y,z directions, which is not the case for an anisotropic scaling.

Andersen barostat. The Andersen barostat [6] is conceptually closer to the Nosé–Hoover thermostat rather than the Andersen thermostat. The Andersen barostat samples in the (NPH) ensemble. It involves a coupling of the system to an external variable V, the volume of the box. The coupling resembles the action of a piston on the system. The coordinates r_i and velocities v_i are being rescaled, so that the new coordinates s_i and velocities $\dot{s}_i$ are $s_i = \frac{r_i}{V^{1/3}}$ and $\dot{s}_i = \frac{v_i}{V^{1/3}}$, respectively and define a new Lagrangian with the new variable Q. This variable corresponds to the coordinate of the piston and its fluctuations can be controlled in the simulations through a variable for the mass of the piston. In this way, a desired pressure can be applied to the system.

3.3 Integration schemes

Newton's equations of motion ($\vec{F}_i(\vec{r}_i) = m_i \frac{d^2\vec{r}_i}{dt^2}$) can be solved analytically only for $N < 3$, where N is the number of particles in the system. For $N > 3$ the equations can be solved only numerically by numerical integration. Representative examples of the numerous integration algorithms are presented here. Typically, the integration of the equations of motion is obtained by a Taylor expansion (the so-called Verlet-type-algorithms). Assuming, that $\vec{r}_i$, $\vec{v}_i$, $\vec{F}_i$ are the position, velocity, and force acting on particle i, Δt is the time-step in the simulations, and t the time, the Taylor expansion of the equations of motions, up to orders of three and two, would be given through:

$$\vec{r}_i(t + \Delta t) = \vec{r}_i(t) + \Delta t \cdot \vec{v}_i(t) + \frac{(\Delta t)^2}{2m_i}\vec{F}_i(t) + \mathcal{O}\left((\Delta t)^3\right) \qquad (3.10a)$$

$$\vec{v}_i(t + \Delta t) = \vec{v}_i(t) + \frac{\Delta t}{m_i}\vec{F}_i(t) + \mathcal{O}\left((\Delta t)^2\right) \qquad (3.10b)$$

Often integration schemes can be very efficient, but are not optimum for simulations of physical systems. One has to have in mind the following criteria, before choosing an integration scheme: (a) **accuracy** denoting to which power of the time step the numerical trajectory will deviate from the exact trajectory after one integration step. (b) **energy conservation** along the exact trajectory is a result of the time-translation invariance of the Hamiltonian. The energy of the numerical trajectory will drift (a steady increase or decrease) from the exact one and is characterized by the noise (fluctuations on top of the drift). (c) **time-reversablity** is a result of energy conservation and assures that moving back in time would lead to the same trajectories as when moving forward in time. (d) **symplecticity** is related with the preservation of the phase-space in the simulations. It is physically impossible to remove or add regions in the phase-space without any external reason. Accordingly, the integrator should sample the phase-space, but keep it intact.

The simulations involve an approximate modeling of a physical system and can never be exact, thus they are prone to errors which have different sources. Errors related to the integration schemes may result from the numerical integration method or from finite precision arithmetics, which is inherent in computers. For example, the Verlet algorithm is not susceptible to an energy drift in the exact arithmetic, but errors may occur in practice as a result of finite precision arithmetics when particular numbers are rounded off in computers.

Euler. The Euler integration algorithm is the simplest one [11]. The trajectory is calculated according to the Taylor expansions for the particle positions and velocities in equations (3.10). The Euler algorithm is neither symplectic nor time-reversible. It can be used to integrate other equations of motion, like the Boltzmann equation.

Verlet. Within the Verlet algorithm, the equations of motion are solved based on the current positions $\vec{r}_i(t)$ and forces $\vec{F}_i(t)$ and previous positions $\vec{r}_i(t - \Delta t)$ [12]. A Taylor expansion is taken up to order 4 for positions and velocities. By substracting and adding these expansions, the equations for the integration scheme are obtained:

$$\vec{r}_i(t + \Delta t) = 2\vec{r}_i(t) - \vec{r}_i(t - \Delta t) + \frac{(\Delta t)^2}{m_i}\vec{F}_i(t) + \mathcal{O}\left(\Delta t^4\right) \qquad (3.11a)$$

$$\vec{v}_i(t) = \frac{\vec{r}_i(t + \Delta t) - \vec{r}_i(t - \Delta t)}{2\Delta t} + \mathcal{O}\left(\Delta t^3\right) \qquad (3.11b)$$

The integration starts by specifying $\vec{r}(t)$ and $\vec{v}(t)$. Note, that within the Verlet algorithm the velocities are not needed to compute the trajectories (positions), but are useful for calculating observables like the kinetic energy and the instantaneous temperature. The velocities $\vec{v}_i(t)$ are only available once $\vec{r}_i(t + \Delta t)$ has been calculated one time-step later.

The Verlet algorithm is fast and time reversible. It is symplectic as it conserves the volume of the phase space as imposed by the Hamiltonian dynamics. The Verlet algorithm shows a good short-term energy conservation and a small long-term energy drift, while the trajectories are very close to the constant energy hyper-surface in phase space. A strong disadvantage of the algorithm is that the velocities do not enter directly, which makes the simulations at a constant temperature difficult.

Leap-frog. The leap-frog algorithm [13] uses the Euler method to numerically solve differential equations and obtains both particle positions ($\vec{r}_i$) and velocities ($\vec{v}_i$) from readily available quantities. It approximates $\vec{r}_i$ with its value at the midpoint of the time interval, so that

$$\vec{r}_i(t + \Delta t) = \vec{r}_i(t) + \vec{v}_i\left(t + \frac{\Delta t}{2}\right)\Delta t \qquad (3.12a)$$

Similarly, the midpoint rule is taken to propagate the velocities:

$$\vec{v}_i\left(t + \frac{\Delta t}{2}\right) = \vec{v}_i\left(t - \frac{\Delta t}{2}\right) + \frac{\vec{F}_i(t)}{m_i}\Delta t \qquad (3.12b)$$

and the leap-frog algorithm generates positions and velocities shifted by $\frac{\Delta t}{2}$. The velocities are updated at half time-steps and 'leap' ahead the positions. In this way, numerical imprecisions are minimized, but the velocities are still not accessible ad hoc. One must first update the positions, recalculate the forces, and update the velocities. Since only $\vec{r}(0)$, $\vec{v}(0)$ are given as an input, the algorithm starts by first performing a single-half time-step. The leap-frog algorithm is both symplectic and time-reversible.

Velocity-Verlet. The velocity-Verlet integration algorithm [14] is mathematically equivalent to the original Verlet algorithm. Positions, velocities, and forces are obtained at the same time, which is necessary for the calculations of the kinetic, potential, and total energy at the same time-step. In order to formulate this algorithm, one has to split the leap-frog algorithm into half steps. This scheme is equivalent to the leap-frog algorithm. The forces acting on each particle are still only calculated once per time-step. After some elaboration on the equations for the leap-frog algorithm, according to the above, the equations for the velocity-Verlet scheme can be simplified to

$$\vec{v}_i(t + \Delta t) = \vec{v}_i(t) + \frac{\Delta t}{2m_i}\left(\vec{F}_i(t) + \vec{F}_i(t + \Delta t)\right) + \mathcal{O}\left((\Delta t)^3\right) \qquad (3.13a)$$

$$\vec{r}_i(t + \Delta t) = \vec{r}_i(t) + \vec{v}_i(t)\Delta t + \frac{(\Delta t)^2}{m_i}\vec{F}_i(t) + \mathcal{O}\left((\Delta t)^3\right) \qquad (3.13b)$$

The velocity-Verlet algorithm is symplectic and very stable and gives particle positions and velocities at the same time-steps.

Overall, the leap-frog and velocity-Verlet algorithms are less susceptible to round-off errors compared to the Verlet algorithm (comparison of $\mathcal{O}(\Delta t)$ versus $\mathcal{O}(\Delta t^2)$ terms). The former two schemes also conserve the angular momentum. All three integration methods are time-reversible and symplectic. In addition to these, other types of algorithms include higher order terms in the Taylor expansion for the positions, velocities in equations (3.10). Such algorithms are the **Beeman's** algorithm, [15], the **Gear predictor-corrector** [16], the **Forest–Ruth** integrator [17], etc. It is possible to reach a better conservation of energy and longer time-steps, that is, shorter simulation time and less demanding computations. Some integration algorithms are often non-symplectic or require several force calculations per time-step, which is computationally too demanding [18, 19].

3.4 General remarks on MD

MD simulations span a number of spatial and temporal time scales. Based on the continuous methodological development and the increasing computational power, these scales are constantly being extended. For example in biomolecular simulations, typically involving the modeling of biomolecules in a solvent, in 2001 it was possible to simulate about 10^6 atoms for 1 ns. In 2010, a 1 ms simulation of a small protein in water was reported. Within the next decade, the simulation of biomolecular systems for 1 ms should be possible. What hinders the use of very large systems and their simulations for very long times are mainly the computational resources, which are relevant to how fast the simulations can converge.

Convergence

Every simulation of a physical system needs to be converged, either this means that its minimum energy is found or the forces acting on the system are below a tolerance. The simulation should have converged until the desired simulation time has been reached. The time scales within various systems (for example related to dynamics of biomolecules or materials' processes) range from fs to s or longer. As mentioned above, current MD cover up to tens of thousands of ns depending on the system. The question arises whether these time scales are long enough to generate reliable trajectories, properties, etc. Let Q be a property of a system, which needs to be determined, τ_{equil} the equilibration time of the simulation, τ_{relax} the relaxation time of property Q, and τ_{sample} the sampling time. Reliable results are expected when

$$\tau_{\text{equil}} > \tau_{\text{relax}}(Q)$$
$$\tau_{\text{sample}} \ll \tau_{\text{relax}}(Q)$$

Accordingly, the choice of $\tau_{\text{relax}}(Q)$ is of high importance when targeting the property Q. The three following details should be considered: (i) in equilibrium simulations it is necessary to monitor the time dependence of the property $Q(t)$, its average value $\langle Q(t) \rangle$, its fluctuations $\langle (Q(t) - \langle Q(t) \rangle)^2 \rangle^{1/2}$ or the relevant autocorrelation function $\langle Q(t')Q(t' + t) \rangle$. The time $\tau_{\text{relax}}(Q)$ should better be defined from the decay time of the autocorrelation function or the simulation build-up rate of the trajectory/property averages. (ii) When the simulation starts from a non-equilibrium initial state, $\tau_{\text{relax}}(Q)$ is given from the rate of relaxation of $Q(t)$ towards the equilibrium and is measured over many non-equilibrium trajectories. (iii) In cases for which MD simulations starting from a different initial condition do not converge to the same averages for Q, then $\tau_{\text{relax}}(Q)$ is longer than the simulation time and the simulations should run over a longer time in order to obtain reliable results for the property Q.

Basic problems of classical modeling

There are some important aspects which inhibit the production of reliable physical properties when performing classical simulations, such as MD. These need to be carefully taken into consideration in advance and are the following: (a) **force-fields/ potentials** are parameterized based on experimental data or more accurate quantum-mechanical simulations. There is a wide range of atoms/molecules resulting in a zoo of different potential functionals and their parameters. Entropic effects are not easily captured and often very small (free) energy differences and many interactions are involved in the modeling. (b) **search problem**: reaching convergence can be tedious without a good search/integration scheme, especially when the potential surface of the physical system is too complex (see chapter 8). In addition, alleviating factors (e.g. the state of the system contains much fewer conformations significantly populated at equilibrium than the possible conformations for the system) and aggregating factors (inclusion of degrees of freedom in sampling, which are necessary for good results) need to be considered. (c) **ensemble problem**: averaging or non-linear averaging, as well as entropy often make the definition of the ensemble not straightforward. (d) **experimental problem**: often insufficient experimental data or data with insufficient accuracy exist, while averaging the existing experimental data also enhances the difficulty of mapping the simulation data to experimental ones or the other way around.

References

[1] Frenkel D and Smit B 2001 *Understanding Molecular Simulation: From Algorithms to Applications* vol 1 (San Diego, CA: Academic)

[2] Allen M P and Tildesley D J 1989 *Computer Simulation of Liquids* (Oxford: Oxford University Press)

[3] Rapaport D C 2004 *The Art of Molecular Dynamics Simulation* (Cambridge: Cambridge University Press)

[4] Hünenberger P H 2005 Thermostat algorithms for molecular dynamics simulations *Advanced Computer Simulation* (Berlin: Springer) pp 105–49

[5] Berendsen H *et al* 1984 Molecular dynamics with coupling to an external bath *J. Chem. Phys.* **81** 3684–90

[6] Andersen H C 1980 Molecular dynamics simulations at constant pressure and/or temperature *J. Chem. Phys.* **72** 2384–93

[7] Nosé S 1984 A unified formulation of the constant temperature molecular dynamics methods *J. Chem. Phys.* **81** 511–9

[8] Hoover W G 1985 Canonical dynamics: equilibrium phase-space distributions *Phys. Rev.* A **31** 1695

[9] Parrinello M and Rahman A 1981 Polymorphic transitions in single crystals: A new molecular dynamics method *J. Appl. Phys.* **52** 7182–90

[10] Martyna G J, Tobias D J and Klein M L 1994 Constant pressure molecular dynamics algorithms *J. Appl. Phys.* **101** 4177–89

[11] Hamming R 2012 *Numerical Methods for Scientists and Engineers* (New York: Dover)

[12] Verlet L 1967 Computer experiments on classical fluids. i. Thermodynamical properties of Lennard-Jones molecules *Phys. Rev.* **159** 98

[13] Van Gunsteren W F and Berendsen H J C 1988 A leap-frog algorithm for stochastic dynamics *Mol. Simul.* **1** 173–85

[14] Swope W C, Andersen H C, Berens P H and Wilson K R 1982 A computer simulation method for the calculation of equilibrium constants for the formation of physical clusters of molecules: application to small water clusters *J. Chem. Phys.* **76** 637–49

[15] Beeman D 1976 Some multistep methods for use in molecular dynamics calculations *J. Comput. Phys.* **20** 130–9

[16] Gear C W 1971 *Numerical Initial Value Problems in Ordinary Differential equations* (Englewood Cliffs, NJ: Prentice Hall)

[17] Forest E and Ruth R D 1990 Fourth-order symplectic integration *Physica D* **43** 105–17

[18] Leimkuhler B and Reich S 2004 *Simulating Hamiltonian Dynamics* vol 14 (Cambridge: Cambridge University Press)

[19] Hairer E, Lubich C and Wanner G 2006 *Geometric Numerical Integration: Structure-Preserving algorithms for Ordinary Differential equations* vol 31 (Berlin: Springer)

Chapter 4

Classical potentials and force fields

In simulations of physical systems a key aspect is to model adequately the interactions between the particles of the system. Ongoing work is invested in developing classical inter-particle potentials. For this, three are the most important aspects: **accuracy**, **transferability** and **efficiency**. The potentials need to be accurate enough for the physical systems they have been developed for. They need to be transferable, that is, able to efficiently model additional systems. In addition, classical potentials should have simple mathematical forms in order to be easily implemented in a computational code and not increase the computational time (i.e. destroy the scaling of the respective simulation method). A potential should be able to model sufficiently the different types of bonding that can occur in a physical system. The particles of such a system can interact through bonds, ranging from strong covalent bonds up to non-directed long-range interactions [1].

The development of classical potentials involves a lot of work and testing and can be nailed down into three aspects: (a) an assumption for the functional form of the potential is made and parameters are chosen to reproduce a set of available experimental data. For this, empirical potential forms, such as the known Morse, Lennard-Jones, etc, do exist and can be used. (b) Quantum mechanical arguments are taken to derive analytical semi-empirical potentials, such as the bond-order potentials of Tersoff and Brenner, the embedded atom method (EAM), etc. (c) Direct electronic-structure calculations of forces are performed through *ab initio* MD, such as Car-Parrinello, which are then translated into potential functions. 'Good' potentials are usually derived by combining QM simulations and experimental data to fit to different properties of a system. For solid-state (materials) systems, the interactions are modeled typically through pair-wise potentials. For more complex simulations such as for biomolecules in a solvent, more detailed and complex potentials, the force fields (FFs) are taken and involve a consistent set of parameters. Changing a subset of parameters by taking these from another FF may introduce inconsistencies.

4.1 Classical pair potentials

In classical simulations, the inter-particle interactions are approximated through classical potentials between particle centers ('effective potentials'). The use of these potentials allows classical simulations also for millions of particles. Typically, their potentials depend only on the inter-particle distance $\vec{R}$, as sketched in figure 4.1. The corresponding potential has a repulsive (positive) and an attractive (negative) part. This form corresponds to attraction of the particles when they come close enough and repulsion when these are too close. When the system is in equilibrium the particles are at a distance R_{min}, which corresponds to the minimum of their interaction potential (equation in figure 4.1).

In the simulations, the total interaction (or total energy) of the system is given through a sum over all inter-particle interactions. The advantage of using pair potentials is the relative fast computation. For an N particle system with proper optimization (cutoffs, etc), the scaling is $\mathcal{O}(N)$. On the other hand, the pair potentials are often a too fair description of the real interactions. They ignore orbital interaction and often polarizability and neglect multi-body effects, as not all interactions can be represented through pair potentials. Usually chemical reactions are not possible to be directly described through simple classical potentials, as bond-breaking is based on electronic features. Exceptions such as reactive force fields [2] have recently been developed.

In general, a system of N interacting particles described through empirical (classical) potentials has a total energy given through

$$U(\vec{r}_1,...,\vec{r}_N) = \sum_i U_1(\vec{r}_i) + \sum_i \sum_{j>i} U_2(\vec{r}_i, \vec{r}_j) + \sum_i \sum_{j>i} \sum_{k>j} U_3(\vec{r}_i, \vec{r}_j, \vec{r}_k) + \cdots \qquad (4.1)$$

where U_1 is a one-body term due to external field or boundary conditions, U_2 is a two-body term (pair potential) for which the interaction depends on the spacing of two atoms and is not affected by the presence of other atoms, and U_3 is a three body term, which maps the modified interaction between a pair of atoms due to the presence of a third atom. Additional terms can always be added. In this way, more details of the system are involved increasing, though, the computational load.

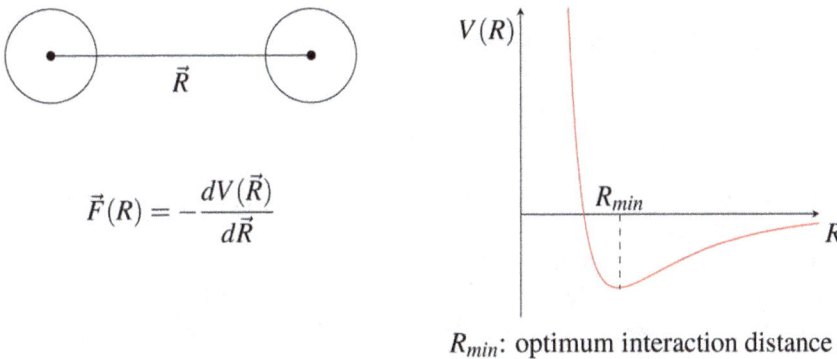

$$\vec{F}(R) = -\frac{dV(\vec{R})}{d\vec{R}}$$

R_{min}: optimum interaction distance

Figure 4.1. Left: Interaction of two particles at distance $\vec{R}$ on which a force $\vec{F}(\vec{R})$ is acting. Right: $\vec{F}$ is defined through the inter-particle potential $V(R)$, with R_{min} the equilibrium distance.

4.1.1 Simple pair potentials

The simplest many-body systems to be captured by a simple pair potential are the ideal gas of non-interacting particles (not very relevant to real systems) and the real gas in which particles have a size, a short range repulsion and a possible short range attraction. Simple model potentials can be used for approaching in a very simple way the inter-particle interactions (controlled by their distance r_{ij}) in these systems, such as the hard sphere potential

$$V^{\mathrm{HS}}(r_{ij}) = \begin{cases} \infty, & r_{ij} < \sigma \\ 0, & r_{ij} \geqslant \sigma \end{cases}$$

the square-well potential

$$V^{\mathrm{SW}}(r_{ij}) = \begin{cases} 0, & r_{ij} < \sigma_1 \\ -\varepsilon, & \sigma_1 \leqslant r_{ij} \leqslant \sigma_2 \\ 0, & r_{ij} > \sigma_2 \end{cases}$$

the soft-sphere potential

$$V^{\mathrm{SS}}(r_{ij}) = \varepsilon \left(\frac{\sigma}{r_{ij}}\right)^{\nu} = \alpha r_{ij}^{-\nu}$$

the harmonic potential

$$V^{\mathrm{h}}(r_{ij}) = a_0 + \frac{1}{2}k(r_{ij} - r_0)^2$$

the Buckingham potential

$$V^{\mathrm{B}}(r_{ij}) = C\left(\frac{\sigma}{r_{ij}}\right)^6 - Ae^{-\sigma r_{ij}}$$

The Morse

$$V^{\mathrm{Morse}}(r_{ij}) = D\left(e^{-2a(r_{ij}-r_0)} - 2e^{-a(r_{ij}-r_0)}\right)$$

and the Lennard-Jones potential

$$V^{\mathrm{LJ}}(r_{ij}) = 4\varepsilon\left[\left(\frac{\sigma}{r_{ij}}\right)^{12} - \left(\frac{\sigma}{r_{ij}}\right)^6\right]$$

are often used to model long-range van der Waals interactions. The Lennard-Jones potential is sketched in figure 4.2 Its r^{-6} term maps the (dipole–dipole, dipole-induced dipole, and London) attraction at longer ranges, while the r^{-12} term is based on Pauli's exclusion principle due to overlapping electron orbitals. One of the early more complicated three-body potentials is the Stillinger–Weber potential [3], which has transferability issues as the three-body term defines only one equilibrium configuration.

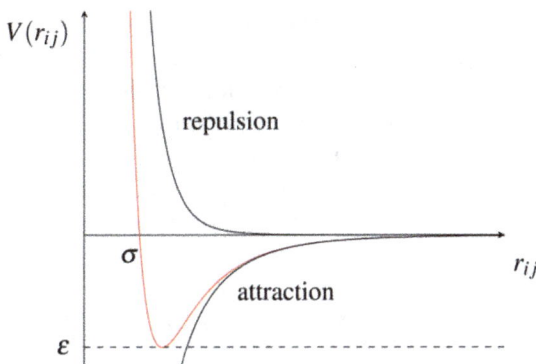

Figure 4.2. The Lennard-Jones potential (red line) decomposed into a repulsive and attractive part (black lines).

Note, that all the unknowns apart from the inter-particle distance (r_{ij}) in the above equations are parameters which need to be tuned for each chemical element. Accordingly, simple pair potentials have limited transferability. Parameters used for molecules cannot always be used for crystals of the same elements and might also not capture the difference between crystal lattices. Most of these issues can be better approached with the use of bond-order potentials or reactive FFs, which will be discussed later in this chapter. Pair potentials, such as the ones above, are computationally very efficient. However, an important point not addressed in these models is the differences between the bonds and the dependence of the environment in binding.

4.1.2 Bond-order potentials

A feature which needs to be added in the pair potentials is the coordination of an atom (i.e. the number of its nearest neighbors), often recast in the form of the bond-order. The bond-order is the number of chemical bonds between a pair of atoms and does not need to be an integer. The bond-order accounts for the environment of an atom and is the essential ingredient of **bond-order pair potentials**. In a conventional potential like Morse or Lennard-Jones, the bond length between the particles solely defines their interaction energy. In bond-order potentials, the bond length, through the Pauling principle is related to the bond-order and together define the total energy.

Examples of bond-order potentials are the Tersoff [4] and the Brenner potentials [5], both based on the Abell theory [6] and the environment-dependent interatomic potential (EDIP) [7]. The Tersoff potential is very efficient in modeling simple covalent single- and multicomponent systems. It shows problems related to over-binding of radicals (overestimation of the density for specific hybridizations) and conjugacy (i.e., overlap of two p-orbitals, bridging the interjacent single bonds). The Brenner potential corrects for these and also includes non-local effects. Such a potential is also known as a reactive bond-order (REBO) potential. Its shortcomings are the absence of non-bonding interactions and torsional parameters for hindered rotations about single bonds. An additional improvement is the adaptive inter-molecular reactive bond-order potential (AIREBO) [8], which includes covalent and non-bonded interactions. Both repulsive and attractive pair interaction functions in REBO are modified to fit bond properties, while long-range atomic interactions and single-bond torsional interactions are included. It is also quite transferable (compared to REBO) and is an empirical potential capable of simulating chemical reactions in different environments.

The AIREBO potential depends on the functional form of a REBO potential (E^{REBO}) and its total energy is defined through $E = E^{\text{REBO}} + E^{LJ} + E^{\text{tors}}$. The Lennard-Jones term E^{LJ} for the non-bonded interactions can be switched off depending on the distance and the bonding environment and includes a bond-order term. The torsional term E^{tors} is given through

$$E^{\text{tors}} = \frac{1}{2} \sum_i \sum_{j \neq i} \sum_{k \neq i,j} \sum_{l \neq i,j,k} w_{ij}(r_{ij}) w_{jk}(r_{jk})_{kl}(r_{kl}) V^{\text{tors}}(\omega_{ijkl}) \qquad (4.2)$$

where ω_{ijkl} is the dihedral angle and $w_{ij}(r_{ij})$ are bond weights ensuring that the torsional energy associated with a given dihedral angle will be removed smoothly when any of the constituent bonds are broken.

4.2 Multi-body reactive force fields

The potentials reviewed up to this point can be easily implemented in a computational code and better model covalently bonded systems. Their transferability is not always very good and they poorly (apart from the AIREBO potential) capture non-bonded interactions, bond-breaking and chemical reactions. To this end, more efficient and transferable potentials are being developed.

4.2.1 Reactive force field (ReaxFF)

The reactive force field (ReaxFF) [2] is a molecular model to describe chemical reactions. It can model a continuous energy landscape during reactions and involves certain element types. The potential is computationally efficient as it includes finite-range interactions being able to treat large systems ($\approx 10^4$ atoms). A typical form for the total energy of a system through a ReaxFF is given by:

$$E_{\text{system}} = \underbrace{E_{\text{bond}} + E_{\text{vdW}} + E_{\text{Coul}}}_{2-\text{body}} + \underbrace{E_{\text{val,angle}}}_{3-\text{body}} + \underbrace{E_{\text{tors}}}_{4-\text{body}} + \underbrace{E_{\text{over}}}_{\text{multi}-\text{body}} \qquad (4.3)$$
$$+ E_{\text{under}} + E_{\text{pen}} + E_{\text{conj}}$$

In this expression, E_{bond} is the bond energy depending on the bond-order, E_{vdW} is the dispersion energy (non-bonded van der Waals interactions), E_{Coul} are the electronic interactions, $E_{\text{val,angle}}$ is the angle strain term, E_{tors} the torsional energy, E_{over} the over-coordination energy imposing a penalty for over-coordinated atoms, E_{under} the under-coordination energy, which considers the energy contribution for resonances of n electrons between attached under-coordinated atomic centers. E_{pen} is a penalty term for 'allene'-type of molecules ($H_2C=C=CH_2$)[1] and E_{conj} is a term accounting for conjugation effects to the molecular energy.

---

[1] It is a penalty in order to produce stability of a system with two double bonds sharing an atom in a valence angle.

All terms in equation (4.3) describe individual chemical bonds and are given as expressions of the bond-order. No atom type (e.g. hybridization) is accounted for, only the element type is considered. The fundamental assumption of the ReaxFF is that the bond-order (BO) between a pair of atoms is obtained directly from the inter-atomic distance and involves contributions from σ, π and $\pi-\pi$ bonds ($BO_{ij} = BO_{ij}^{\sigma} + BO_{ij}^{\pi} + BO_{ij}^{\pi\pi}$). Polarization effects can be accounted for due to the geometry dependent charge calculation scheme with reactive force fields. One of the main strengths of ReaxFF is that it describes properly bond dissociation and can efficiently model chemical reactions. However, for specific systems, not all terms in equation (4.3) are relevant. The following descriptions can be assumed depending on the system type:

covalent: $E_{system} = E_{bond} + E_{over} + E_{val} + E_{tors} + E_{VdWaals} + E_{Coulomb}$

metal alloys: $E_{system} = E_{bond} + E_{over} + E_{VdWaals} + E_{Coulomb}$

metals: $E_{system} = E_{bond} + E_{over} + E_{VdWaals}$

ionic materials: $E_{system} = E_{Coulomb} + E_{VdWaals}$

noble gases: $E_{system} = E_{VdWaals}$

4.3 Force fields for biomolecules

Solid-state systems are not so flexible and are often governed by strong two-body interactions, compared to biomolecular systems which are much more flexible, include typically non-bonded interactions and need a more complex description. This energetic description is overall known as the force fields (FFs). Biomolecules resemble 'legos', as they consist of distinct units (e.g. DNA is a sequence of nucleobases, proteins consist of amino acids, etc). It can be assumed, that the atom–atom interactions can be described by the same potential with different parameters, but should always be tested first.

In a biomolecular force-field, **covalent** interactions between atoms (bond-stretching (E_{bond}), bond-angle (E_{angl}), bending, improper/proper dihedral-angle torsion E_{tors}), and **non-bonded** interactions between atoms or units in different or the same molecule (separated by more than two or three covalent bonds) should be properly described. Typically, the most important non-bonded interactions are van der Waals (E_{vdW}) and Coulomb (E_{Cb}). The respective total energy would be expressed like

$$E_{tot} = \underbrace{E_{bonded}}_{=E_{bond}+E_{ang}+E_{tors}} + \underbrace{E_{non-bonded}}_{+E_{vdW}+E_{Cb}} \qquad (4.4)$$

Common expressions for the different interactions, as well as a sketch of the parameters involved, are given below.

$$E_{bond} = \sum_{bonds} S_{ij}(r_{ij} - r_{ij}^0)$$

distance r_{ij} ; covalent bond with equilibrium-bond stretching term r_{ij}^0

$$E_{ang} = \sum_{i,j,k} K_{ijk}(\cos\theta_{ijk} - \cos\theta_{ijk}^0)$$

covalent bond-angle ; equilibrium angle θ_{ijk}^0 bond bending term

$$E_{tor} = \sum_{i,j,k,l}\sum_n \frac{V_{nijkl}}{2}\left[1 \pm \cos\left(n\tau_{ijkl}\right)\right]$$

torsional angle τ_{ijkl}; dihedral angles-torsion term

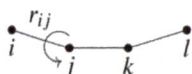

$$E_{vdW} = \sum_{i,j}\left(-\frac{A_{ij}}{r_{ij}^6} + \frac{B_{ij}}{r_{ij}^{12}}\right)$$

van der Waals interaction through a Lennard-Jones potential

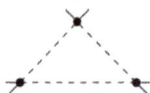

$$E_{Cb} = \sum_{i,j}\left(\frac{q_i q_j}{\varepsilon(r_{ij})r_{ij}}\right)$$

Coulomb interactions between partial charges q_i ; q_j $\varepsilon(r_{ij})$: distance-dependent dielectric constant

These expressions include distances and angles and a number of parameters, which need to be tuned. The FF parametrization is based on experimental (x-rays, IR, etc) and QM data. The structural data are typically used to get the bond parameters. Spectroscopic data deliver vibrational forces and the angle parameters. Dielectric data, that is, dielectric permitivities can lead to the charges, while transport data are related to diffusion coefficients and can be used to obtain the non-bonded van der Waals parameters. The fitting parameters are not unique, but take into account the specific system and fitted properties for which the FF should be applicable. Regarding the applicability and efficiency of an FF, often it cannot be applied to a variety of molecules it was not developed for. In addition, the FF parameters are consistent with specific solute molecules and solvent and cannot be used universally. Terms additional to those in equation (4.4) are often included, such as improper dihedral angles (torsion in terms of geometry, not chemistry) and others.

The most commonly used biomolecular FFs are the CHARMM (Chemistry at HARvard Molecular Mechanics) [9] and the PARM (or AMBER) [10], GROMOS (Groningen Molecular Simulation) [11] and OPLSAA [12] FFs, but additional ones can be found in the literature. Overall, the parameter development for novel molecules within the FF approach is achieved by following specific (published) protocols. The major computational effort is found in the evaluation of non-bonded interactions. Taking into account that a bio-molecular system is usually in a condensed phase, data from the condensed phases are used when possible for the benchmark simulations and the parameterization. The transferability of the FF can be maximized by using only data for small molecules. When using data from large molecules, the properties of groups of atoms may depend on the particular environment. The FF is tested by applying it to systems containing different, larger molecules in the condensed phase and the simulated properties are then compared with available experimental data.

4.4 Embedded atom method (EAM)

Types of systems different than what was reviewed up to this point are metallic systems. In these, the metallic bond involves electrostatic attractive forces between delocalized (conduction) electrons gathered in an electron cloud and the positively charged metal ions. The conduction electrons distribute their density equally over all atoms functioning as neutral (non-charged) entities. The conduction electrons move between the nuclei generating a binding force to hold atoms together. This is known as the 'electron gas model', within which the positive ions move in an electron sea [13]. Metallic systems involve mostly **non-directional** bonding and the dependence of strength of individual bonds on the local environment is especially important for the simulation of surfaces and defects. Classical pair potentials cannot adequately describe metallic systems, either because they do not account for the environment or for the directional nature of bonds. These potentials are also inefficient in dealing with covalent contributions (d-orbitals) of transition metals. A good model potential for metallic systems is the embedded atom method (EAM) [14, 15].

EAM can efficiently treat metallic systems by including electron density effects. Within EAM the description of bonding in metallic systems is based on the concept of local density. The environment of an atom, i.e. the electron density imposed by other atoms, influences the bond strength. Each atom features a certain distribution of the electron density. The local or atomic electron density $\rho_i(r_{ij}) = \sum_{j \neq i} \Pi j(\vec{r}_{ij})$ is the contribution to the electron density of atom i due to the electron density of atom j evaluated at distance r_{ij} and is based on a pair potential $\Pi j(\vec{r}_{ij})$. The total energy of a metallic system of N atoms is defined as

$$E_{\text{tot}} = \sum_i E_i \tag{4.5a}$$

$$E_i = \frac{1}{2} \sum_{j \neq i} \phi_{ij}(r_{ij}) + F_i(\rho_i) \tag{4.5b}$$

where $F_i(\rho_i)$ is the embedding energy as a function of electron density and the pair potential between particles i and j. It is a measure of how locally the electron contributes to the potential energy. The two components of the atomic energies are sketched in figure 4.3.

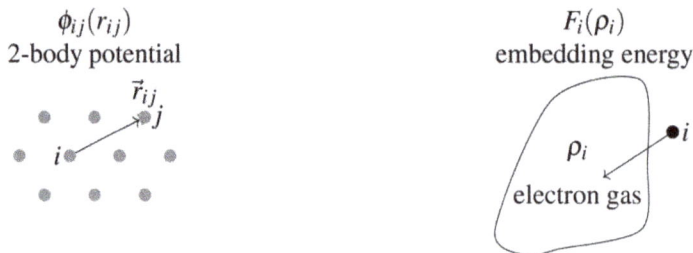

$$\phi_{ij}(r_{ij}) \qquad\qquad\qquad F_i(\rho_i)$$

2-body potential $\qquad\qquad\qquad$ embedding energy

$\vec{r}_{ij}$

ρ_i

electron gas

Figure 4.3. A sketch of the two body-potential $\phi_{ij}(r_{ij})$ and the embedding energy term $F_i(\rho_i)$.

The exact functional forms of F, Π, ϕ depend on the system. The embedding energy $F(\rho)$ is the energy associated with placing an atom in the electron environment described by ρ, while the pair potential ϕ describes electrostatic contributions. $F(\rho)$ is a measure of the number of available states to delocalize into (inherently a many-body effect). Regarding the technical characteristics of EAM potentials, these are numerically efficient (simulations can involve up to $\approx 10^9$ particles). EAM is currently the state-of-the-art approach to model metals with atomistic simulations and involves environment dependent pair potentials. However, the lack of explicit three-body terms makes use of EAM potentials for metals where covalent effects are important (e.g. transition metals) quite challenging.

4.5 Water models

Another system, highly relevant to simulations including a solvent is water. This molecule is very simple, two H atoms are bonded to a single O atom in a certain geometry. For water, specific models have been developed. The first attempt to develop a water model was made in 1979 [16]. Many water models try to take into account its V-shape with a molecular diameter of about 2.75 Å and the frequent fact that the three atoms do not stay together as the hydrogen atoms are constantly being exchanged between water molecules due to (de)protonation. Although simple in structure, it possesses complex hydrogen bonding interactions with its neighboring molecules and is a network former. van der Waals forces are also involved in the interactions. Water molecules in liquid water are all non-equivalent and differ in their molecular orbitals, precise geometry, and molecular vibrations due to the hydrogen bonding characteristics, which are influenced by arrangements of surrounding water molecules.

In the classical description, there is no winner among the water models. Each of these, rather matches certain limited properties of water. Here, the basic ones are reviewed. The water models attempt to optimize the three basic parameters: number (and/or position are required) of charges, interactions, and dipole moment. The more fitting parameters required for the model, the better the fit. Some models show a lack of robustness due to their sensitivity to the precise model parameters, system size or simulation method. Some are polarizable, others reproduce average structures. Two large groups of water models exist: (i) explicit and (ii) implicit water models. The first include the atomistic structure of the water molecule, while the second account for the water environment through continuous models.

4.5.1 Explicit water models

Three main water model types exist in explicit solvation: (i) the **rigid** models in which the atoms have fixed positions; only non-bonded interactions are taken into account and the parameters are matched to the known geometry of water. (ii) The **flexible** models assume atoms on 'springs', include bond stretching and angle bending, and reproduce the vibration spectra of water. (iii) The **polarizable** models include explicit polarization terms enhancing the ability of the model to reproduce water in different phases and the interactions between these. In reality, water is flexible and polarizable. Models which do not include these properties may have issues with predictability, which typically decreases as the temperature is lowered. Examples of polarizable water models are the SPC/FQ or TIP4P/FQ [17–18]. The latter being computationally only 1.1 times slower than the corresponding rigid model. Typically, the parameters of the water models are determined through *ab initio* calculations of dimers, trimers or higher order clusters or are empirically (using the Lennard-Jones potential) developed to reproduce experimental data.

Rigid water models

Within the rigid water models, the molecule interacts via non-bonded (Lennard-Jones) interactions with other molecules. Some of these were developed within a specific FF and are often adapted to other FFs. For example, the GROMOS FF has been developed for SPC, SPC/E, AMBER for TIP3P, CHARMM for a modified TIP3P, while the OPLS force field has been developed with TIP4P and TIP5P. Accordingly, use of another water model with a specific FF could most probably lead to inconsistent or unreliable results and should be first tested. The different types of rigid water models depend on the number of parameters needed to define the model:

- **3-site:** each atom gets a point charge assigned and has its own Lennard-Jones parameters. Common 3-site models are the SPC (single point charge) [19], SPC/E [20], the flexible SPC [20], and TIP3P [21].
- **4-site:** the negative charge is assigned to a dummy atom instead to the position of the O. This improves the electrostatic distribution around the water molecule. The most common 4-site models are TIP4P [21] or flexible TIP4PF [22].
- **5-site:** the negative charges are split and sit on the lone pairs[2] of oxygen with a tetrahedral-like geometry. A representative model is the TIP5P model [23].
- **6-site:** combines all characteristics of the 4-site and 5-site models [24].

Another issue related to transferability is the fact that the solid phase of water exhibits one of the most complex phase diagrams with 13 different (known) solid structures. From the simple water models (SPC, SPC/E, TIP3P, TIP4P and TIP5P) only TIP4P provides a qualitatively correct phase diagram on water. For issues like this, other models have been developed, like the F3C [25] or DEC, which employs

---

[2] A lone pair is a valence electron pair without bonding or sharing a bond with other atoms.

diffuse charges, additionally to the usual point charges [26]. The TIP4P/FLEX describes well the absorption spectra for liquid water [27]. The MB model reproduces the features of water in 2-D systems and is rather for educational purposes than for real simulations [28]. Coarse-grained, one- and two-site models, where each site represents a number of water molecules are also known [29].

4.5.2 Implicit water models

Another large family of water models are the implicit water models [30–32], which do not account for the atomistic details the molecule, as in the explicit water models. Implicit water models represent the solvent and counter-ions as a continuous medium and solvent molecules are not explicitly taken into account. The continuous medium has specific solvation and dielectric properties. Simulations with implicit water are typically faster than explicit simulations and are very good approximations when the distribution of individual water molecules in the solvent–solute interface is not of interest. Within implicit models, errors in statistical averaging arising from incomplete sampling of solvent conformations can occur. The most common implicit water models are the solvent accessible surface area models, continuum electrostatics (e.g. Poisson–Boltzmann equation), and the generalized Born (GB) models.

Solvent accessible surface area (SASA)
The solvent accessible surface area (SASA) is a concept based on experimental relations between the Gibbs free energy of transfer and the surface area of the solute. This approach is different than electrostatic methods, which include only the enthalpic component of the free energy [33]. SASA uses spheres to model atoms and solvent molecules. The free energy of solvation (ΔG_{solv}) can be obtained through a summation over all atoms in the simulations

$$\Delta G_{solv} = \sum_i \sigma_i ASA_i \qquad (4.6)$$

where, σ_i is an atom-specific surface tension parameter estimated from empirical hydration free energies of compounds in water. ASA_i is the surface area of atom i, which is accessible to the solvent. This surface is generated by the center of a sphere/ probe rolling on the van der Waals surface of the simulated atoms. Drawbacks of the SASA scheme are that it can overestimate solvation free energies and ignores the electrostatic effects of the solvent. No interactions between the solvent and polar atoms in the interior of the molecule are considered. This can be a severe restriction when polar groups are of importance.

An example of a SASA algorithm is *the Shrake–Rupley algorithm*, which uses a discretized molecular surface on which 92 points are distributed uniformly along a sphere centered at each atom with a chosen radius [34]. The ASA_i is estimated by calculating the proportion of points that can be contacted by a solvent molecule not intersecting other atoms. Other methods use polyhedra to approximate the surface. The *linear combination of pairwise overlaps* uses an analytical expression based on an inclusion–exclusion-like formula [35]. The expression includes the surface area of an atom, the surface area of a sphere of this atom in another sphere, sums over the neighbors of an atom, and uses least square-regression.

Continuum electrostatic models
The continuum electrostatics models take into account the effects of a solvent on the electrostatics of a molecule. The simplest approach is the use of a distance-dependent dielectric function to mimic the electrostatic screening due to the presence of the solvent. The method ignores the fact that the solute molecule influences the distribution of solvent molecules[3].

The Poisson–Boltzmann equation (PBE) is used in these models [36]:

$$\nabla\left[\varepsilon(\vec{r})\nabla\phi(\vec{r})\right] = -4\pi\rho_{\text{solute}}(\vec{r}) - 4\pi\sum_i q_i c_i e^{-\tilde{V}_i(\vec{r})/k_{\text{B}}T} \qquad (4.7)$$

where $\varepsilon(\vec{r})$ is the distance-dependent dielectric function, ρ_{solute} is the charge distribution of the solute, $\tilde{V}_i(\vec{r})$ the potential of mean force, $k_{\text{B}}T$ the Boltzmann temperature, q_i the charge of the ith ion, and c_i its concentration. Often, in simulations the linearized Poisson–Boltzmann equation (LPBE) scheme is used [37]. The purpose of linearization is to solve PBE more efficiently[4] and can be done if $q_i\phi(\vec{r}) \ll k_{\text{B}}T$. The linearization leads to a solution, which considers a single spherical ion in an electrolyte with a radius, a uniform surface charge, and constant interior and solvent permittivities. For numerically solving PBE or LPBE, finite-difference, finite-element and boundary-element approaches have been developed [38]. The solutions need to be smooth in space and time in order to be efficiently used in MD. The boundary conditions should be accurately defined and the dielectric function should change rapidly at the solvent–solute interface.

Generalized Born model
The GB model is an approximation to the Poisson–Boltzmann equation [39]. The solvation free energy is approximated through:

$$\Delta G_{\text{GB}} = -\left(1 - \frac{1}{\varepsilon_w}\right)\sum_i \frac{q_i^2}{2\alpha_i} - \frac{1}{2}\left(1 - \frac{1}{\varepsilon_w}\right)\sum_{i\neq j}\frac{q_i q_j}{\sqrt{r_{ij}^2 + \alpha_i\alpha_j \exp\left(-r_{ij}^2/4\alpha_i\alpha_j\right)}} \qquad (4.8)$$

where q_i is the particle charge, α_i is the effective Born radius of atom i, and ε_w is the dielectric constant of water. The second term in equation (4.8) is the pairwise contribution of the charges of the molecule to the solvation energy. GB corresponds to solvation in an infinite volume of solvent and is often used with SASA to estimate the non-polar solvation energy (GBSA) [40]. GB models the solute as a set of spheres whose internal dielectric constant differs from that of the external solvent. The

---

[3] Solvent molecules often get excluded from the interior of a solute and counter-ions and polar solvent molecules may condense around the charged atoms.
[4] based on the Debye–Hückel theory.

effective Born radii should be well calculated and denote the distance from the atom to the solute surface. This is an approximation as most of the atoms will not be surrounded by spherical cavities. The effective Born radius is chosen so that $-\left(1 - \frac{1}{\varepsilon_w}\right)\frac{q_i^2}{2\alpha_i}$ is the electrostatic solvation energy of an isolated charge surrounded by a spherical shell that excludes the solvent. The effective Born radii can be estimated numerically, but are often evaluated using an approximate analytical expression [41],

Ad hoc fast and hybrid solvation models
Apart from the implicit water models briefly reviewed above, other strategies, such as ad hoc fast solvation models exist [42]. These perform a calculation of a per-atom solvent accessible surface area. For each group of atom types, a different parameter scales its contribution to the solvation. The Gaussian-shaped solvent exclusion uses a reference free energy of solvation, which corresponds to a suitably chosen small molecule in which a group is fully solvent-exposed [43].

Hybrid solvation models have also been developed [44–46]. These typically include a layer of spheres of water molecules around the solute and model the bulk with implicit solvent. In these models, the viscosity, hydrophobic effect, and hydrogen-bonds with the solvent are not accounted for. The mean electrostatic free energy is estimated without entropic effects due to the solute-imposed constraints on the organization of water. As the viscosity is typically absent, the water molecules impart on the viscosity by randomly colliding and impeding the motion of solutes through their van der Waals repulsion. The average energetic contribution of the hydrogen bonds is in this way reproduced. Overall, a lot of different possibilities occur in modeling a solvent. The model chosen is related mainly to the targeted properties of the simulated system and the solvent features important for these.

References

[1] Pauling L 1960 *The Nature of the Chemical Bond and the Structure of Molecules and Crystals: An Iintroduction to Modern Structural Chemistry* vol 18 (Ithaca, NY: Cornell University Press)
[2] Van Duin A C T, Dasgupta S, Lorant F and Goddard W A 2001 ReaxFF: a reactive force field for hydrocarbons *J. Phys. Chem.* A **105** 9396–409
[3] Stillinger F H and Weber T A 1985 Computer simulation of local order in condensed phases of silicon *Phys. Rev.* B **31** 5262
[4] Tersoff J 1988 New empirical approach for the structure and energy of covalent systems *Phys. Rev.* B **37** 6991
[5] Brenner D W 1990 Empirical potential for hydrocarbons for use in simulating the chemical vapor deposition of diamond films *Phys. Rev.* B **42** 9458
[6] Abell G C 1985 Empirical chemical pseudopotential theory of molecular and metallic bonding *Phys. Rev.* B **31** 6184
[7] Bazant M Z, Kaxiras E and Justo J F 1997 The environment-dependent interatomic potential applied to silicon disordered structures and phase transitions *MRS Proceedings* vol 491 (Cambridge: Cambridge University Press) p 339

[8] Stuart S J, Tutein A B and Harrison J A 2000 A reactive potential for hydrocarbons with intermolecular interactions *J. Chem. Phys.* **112** 6472–86

[9] Brooks B R *et al* 1983 CHARMM: a program for macromolecular energy, minimization, and dynamics calculations *J. Comput. Chem.* **4** 187–217

[10] Cornell W *et al* 1995 A second generation force field for the simulation of proteins, nucleic acids, and organic molecules *J. Am. Chem. Soc.* **117** 5179–97

[11] Scott W R P *et al* 1999 The GROMOS biomolecular simulation program package *J. Phys. Chem.* A **103** 3596–607

[12] Jorgensen W L and Tirado-Rives J 1988 The OPLS [optimized potentials for liquid simulations] potential functions for proteins, energy minimizations for crystals of cyclic peptides and crambin *J. Am. Chem. Soc.* **110** 1657–66

[13] Kittel C 2005 *Introduction to Solid State Physics* (New York: Wiley)

[14] Daw M S and Baskes M I 1984 Embedded-atom method: derivation and application to impurities, surfaces, and other defects in metals *Phys. Rev.* B **29** 6443

[15] Daw M S, Foiles S M and Baskes M I 1993 The embedded-atom method: a review of theory and applications *Mater. Sci. Rep.* **9** 251–310

[16] Barnes P *et al* 1979 Cooperative effects in simulated water *Nature* **282** 459–64

[17] Rick S W, Stuart S J and Berne B J 1994 Dynamical fluctuating charge force fields: Application to liquid water *J. Chem. Phys.* **101** 6141–56

[18] Debye P and Hückel E 1923 The theory of electrolytes. I. Lowering of freezing point and related phenomena *Physikalische Zeitschrift* **24** 185–206

[19] Berendsen H J C *et al* 1981 Interaction models for water in relation to protein hydration *Intermolecular Forces* (Berlin: Springer) pp 331–42

[20] Berendsen H J C, Grigera J R and Straatsma T P 1987 The missing term in effective pair potentials *J. Phys. Chem.* **91** 6269–71

[21] Jorgensen W L and Madura J D 1983 Quantum and statistical mechanical studies of liquids. 25. solvation and conformation of methanol in water *J. Am. Chem. Soc.* **105** 1407–13

[22] Lawrence C P and Skinner J L 2003 Flexible TIP4P model for molecular dynamics simulation of liquid water *Chem. Phys. Lett.* **372** 842–7

[23] Mahoney M W and Jorgensen W L 2000 A five-site model for liquid water and the reproduction of the density anomaly by rigid, nonpolarizable potential functions *J. Chem. Phys.* **112** 8910–22

[24] Nada H and PJM van der Eerden J 2003 An intermolecular potential model for the simulation of ice and water near the melting point: a six-site model of H_2O *J. Chem. Phys.* **118** 7401–13

[25] Levitt M *et al* 1997 Calibration and testing of a water model for simulation of the molecular dynamics of proteins and nucleic acids in solution *J. Phys. Chem.* B **101** 5051–61

[26] Guillot B and Guissani Y 2001 How to build a better pair potential for water *J. Chem. Phys.* **114** 6720–33

[27] Lawrence C P and Skinner J L 2003 Ultrafast infrared spectroscopy probes hydrogen-bonding dynamics in liquid water *Chem. Phys. Lett.* **369** 472–7

[28] Silverstein K A T, Haymet A D J and Dill K A 1999 Molecular model of hydrophobic solvation *J. Chem. Phys.* **111** 8000–9

[29] Izvekov S and Voth G A 2005 Multiscale coarse graining of liquid-state systems *J. Chem. Phys.* **123** 134105

[30] Kleinjung J and Fraternali F 2014 Design and application of implicit solvent models in biomolecular simulations *Curr. Opin. Struct. Biol.* **25** 126–34

[31] Cramer C J and Truhlar D G 1999 Implicit solvation models: equilibria, structure, spectra, and dynamics *Chem. Rev.* **99** 2161–200

[32] Kolar M *et al* 2013 Assessing the accuracy and performance of implicit solvent models for drug molecules: conformational ensemble approaches *J. Phys. Chem.* B **117** 5950–62

[33] Lee B and Richards F M 1971 The interpretation of protein structures: estimation of static accessibility *J. Mol. Biol.* **55** 379–IN4

[34] Shrake A and Rupley J A 1973 Environment and exposure to solvent of protein atoms. lysozyme and insulin *J. Mol. Biol.* **79** 351–71

[35] Weiser J, Shenkin P S and Still W C 1999 Approximate atomic surfaces from linear combinations of pairwise overlaps (lcpo) *J. Comput. Chem.* **20** 217–30

[36] Hansen J- P and McDonald I R 1990 *Theory of Simple Liquids* (Amsterdam: Elsevier)

[37] Tuinier R 2003 Approximate solutions to the Poisson–Boltzmann equation in spherical and cylindrical geometry *J. Colloid Interface Sci.* **258** 45–9

[38] Lu B Z, Zhou Y C, Holst M J and McCammon J A 2008 Recent progress in numerical methods for the Poisson–Boltzmann equation in biophysical applications *Commun. Comput. Phys.* **3** 973–1009

[39] Onufriev A, Case D A and Bashford D 2002 Effective Born radii in the generalized Born approximation: the importance of being perfect *J. Comput. Chem.* **23** 1297–304

[40] Tsui V and Case D A 2000 Theory and applications of the generalized Born solvation model in macromolecular simulations *Biopolymers* **56** 275–91

[41] Hawkins G D, Cramer C J and Truhlar D G 1996 Parametrized models of aqueous free energies of solvation based on pairwise descreening of solute atomic charges from a dielectric medium *J. Phys. Chem.* **100** 19824–39

[42] Wesson L and Eisenberg D 1992 Atomic solvation parameters applied to molecular dynamics of proteins in solution *Protein Sci.* **1** 227–35

[43] Lazaridis T and Karplus M 1999 Effective energy function for proteins in solution *Proteins: Struct., Funct., Bioinform.* **35** 133–52

[44] Lee M S, Salsbury F R and Olson M A 2004 An efficient hybrid explicit/implicit solvent method for biomolecular simulations *J. Comput. Chem.* **25** 1967–78

[45] Keith T A and Frisch M J 1994 Inclusion of explicit solvent molecules in a self-consistent-reaction field model of solvation *Modeling the Hydrogen Bond* vol 569 (Washington, DC: American Chemical Society) pp 22–35

[46] Skyner R E, McDonagh J L, Groom Colin R, Tanja van Mourik and Mitchell J B O 2015 A review of methods for the calculation of solution free energies and the modelling of systems in solution *Phys. Chem. Chem. Phys.* **17** 6174–91

Chapter 5

Mesoscopic particle methods

In the previous chapters, the physical systems were modeled in the lower spatio-temporal scales (see figure 1.2). A link between these lower scales and the macroscopic behavior of a system can be provided through simulations at the mesoscale. These mesoscopic methods involve particles, they are computationally less demanding, and reach much larger and longer spatio/temporal scales. In mesoscopic particle methods, the 'particles' are defined by a set of attributes whose physical meaning depends on the specific level of detail (i.e. scales). These particles can be atoms, molecules (direct simulation of discrete components), volumes of fluid, moving mesh nodes (Lagrangian simulation of continuum equations) depending on the target scale and accuracy. Simulating at the mesoscale is also efficient when hydrodynamics interactions become important and using the 'all-atom' description with explicit solvent particles becomes too expensive.

A scale can be defined as 'mesoscale' if at that length and associated time scale one can assume that the degrees of freedom pertaining to a smaller scale will be in equilibrium when seen from the larger scale. Very common mesoscale methods are the dissipative particle dynamics (DPD) [1], brownian dynamics (BD) [2] including Oseen tensor for hydrodynamic interactions, the lattice Boltzmann method (LBM) [3], stochastic rotational dynamics (SRD) or multi-particles collision dynamics (MPC) [4], etc, and can be applied to fields ranging from the motion of biomolecules in a solution, to dislocations in metals, self-organizing and biomedical materials. Some of these methods are continuum methods (no discrete space is assumed) like the Brownian dynamics, the classical DFT [5], stokesian dynamics [6], DPD or smoothed particle hydrodynamics (SPH) [7]. Others are lattice methods, like the MC lattice chain scheme, the lattice gas automata, the lattice Boltzmann or the lattice director methods. From the above, classical DFT, Brownian dynamics and the MC lattice chain scheme are diffusive methods, while Stokesian dynamics, SPH, DPD, LBM and lattice gas automata are hydrodynamic methods, which can efficiently deal with the presence of a solvent. In the following, some of the continuum and lattice schemes will be briefly reviewed. Additional methods and more details can be found in the literature [3, 8–10].

5.0.1 Simulation of fluids

As discussed in chapter 4, an aqueous environment can be explicitly or implicitly taken into account in atomistic simulations or very explicitly (with all the information on the electrons) in first principles or *ab initio* MD simulations, as discussed in chapter 2. However, it is very difficult to deal computationally with such descriptions in the case of large systems. In addition, it is often not necessary to perform very detailed simulations at those levels, as the desired properties can be obtained through less expensive calculations. For this, fluids are often modeled at the mesoscale. It is essential to solve the correct equations (e.g. Navier–Stokes equation including all the relevant factors) to correctly capture the dynamics and mechanics of the fluid [11–13]. To summarize, modeling of a fluid can be done at three levels:

- Microscopic description: the fluid is assumed as a collection of atoms/ molecules interacting through classical potentials following Newton's equation of motion

$$m_i \ddot{\vec{r_i}} = -\vec{\nabla}_i V(\vec{r_1}, \dots, \vec{r_N}) \tag{5.1}$$

  where m_i is the mass or particle i, $\vec{r_i}$ its position, and $\vec{\nabla}_i V(\vec{r_1}, \dots, \vec{r_N})$ is the force acting on particle i.

- Mesoscopic description: assumes a statistical description of the fluid given by one-particle phase-space distribution functions $f(t, \vec{r}, \vec{v})$, which can define the density

$$\langle n(\vec{r}, t) \rangle = \int f(t, \vec{r}, \vec{v}) \mathrm{d}V \tag{5.2}$$

  and the macroscopic velocity of the fluid

$$\langle v(\vec{r}, t) \rangle = \frac{1}{\langle n \rangle} \int f(t, \vec{r}, \vec{v}) V \ \mathrm{d}V \tag{5.3}$$

  Using kinetic theory, the distribution function f should satisfy the following equation

$$\frac{\mathrm{d}f}{\mathrm{d}t} = \underbrace{\partial_t f + \vec{v} \nabla f}_{\text{free streaming operator}} + \underbrace{C(f)}_{\text{collision operator}} \tag{5.4}$$

  where the free streaming term assumes that the local change in f is due to an independent motion of particles only and is zero in the absence of collisions.

- Continuum picture: at this level, a volume element of the fluid is large enough to neglect the underlying microscopic structure, but is small enough to be treated mathematically as infinitesimal. For example, for the conservation of mass:

$$\oint_\Sigma \rho \underbrace{\vec{v} \ \mathrm{d}\vec{S}}_{\text{flux of mass}} = -\partial_t \underbrace{\int_\Sigma \phi \ \mathrm{d}V}_{\text{mass}} = \int_\Sigma \nabla(\rho \vec{v}) \mathrm{d}V \tag{5.5}$$

and for an infinitesimal volume slice ($\Sigma \to 0$) the continuity equation $\partial_t \rho + \nabla(\rho \vec{v}) = 0$ should hold. For the discretization, approximate equations using suitable discretization techniques and numerical solving need to be taken.

5.1 Continuum methods

The continuum methods involve the level of modeling which is being done in a continuum space. Accordingly, the particles—no matter of what nature—are moving in this continuum space and the respective equations are solved in the continuum.

5.1.1 Brownian dynamics

A very common, conceptually easy to implement and computationally cheap method is brownian dynamics (BD). The Brownian motion of particles is a random walk, a path consisting of random successive steps. It is of a stochastic nature and involves random kicks between particles [2]. The theory involves the formulation of a diffusion equation for Brownian particles in which a diffusion coefficient is related to the mean square displacement of a Brownian particle and could be used to determine measurable physical quantities. The Brownian motion is an experimental evidence for the kinetic theory of gases. Regarding the dynamics of Brownian particles a core question is how far these can travel in a given time interval. For this, classical mechanics is not useful as a Brownian particle undergoes $\sim 10^{21}$ collisions/sec. The collective motion of Brownian particles is considered through a diffusion equation, which in 1-D is expressed as

$$\frac{\partial \rho(x,\ t)}{\partial t} = D\frac{\partial^2 \rho(x,\ t)}{\partial x^2} \tag{5.6}$$

where, $\rho(x,\ t)$ is the density of the Brownian particles at point x and time t and D is their mass diffusion constant. Solving the diffusion equation is in most of the cases difficult because of the external potentials, generic boundary conditions, complex inter-particle interactions, etc. This is the point where simulations can be very useful. It is possible to use an all-atom MD approach which might not be very trivial, as often too many degrees of freedom are involved and only small length and time scales are accessible. An alternative would be a coarse-grained approach through the Langevin equation.

5.1.2 Langevin approach

The Langevin equation describes the motion of a particle in a dissipative medium, where it experiences for example frictional forces. The thermal ('kicking') forces of the surrounding particles are represented by an average time-dependent force and are described by the Langevin equation [14]. It is a stochastic differential equation describing the time evolution of a subset of the degrees of freedom, which are typically collective (macroscopic) variables. These change only slowly compared to the other microscopic variables of the system. The fast (microscopic) variables are responsible for the stochastic nature of the Langevin equation. When a heavy particle is moving through the solvent, it will encounter more solvent particles in the front than in the back. The collisions then with the solvent particles will on average have the effect of a friction force proportional and opposite to the velocity of the heavy particle, suggesting the equation of motion for the heavy particle of mass m:

$$m\frac{d\vec{v}}{dt} = -\gamma\vec{v}(t) + \vec{F}(t) \tag{5.7}$$

where $\gamma(t)$ is the hydrodynamic drag and $\vec{F}(t)$ is the external or systematic force due to other heavy particles, walls, gravitation, etc. When $\vec{F}(t) = 0$, the heavy particle comes to rest, but in reality should perform a Brownian motion. A more realistic model includes a random force $R(t)$ for the frequent collisions with the solvent particles on top of the coarse-grained friction force:

$$m\frac{d\vec{v}}{dt} = -\gamma\vec{v}(t) + \vec{F}(t) + R(t) \tag{5.8}$$

The time correlation should show up in $R(t)$, but is again neglected and $R(t)$ satisfies that the average effect of collisions is already absorbed in the friction term and the expectation value of $R(t)$ vanishes ($\langle R(t)\rangle = 0$). The values of $R(t)$ are also uncorrelated ($\langle R(t)R(t + \tau)\rangle = 0$ for $\tau > 0$).

Two over-simplifications in constructing the Langevin equation are the absence of correlations in the random force and the fact that the frictional force does not depend on the history of the system. The motion of fluid particles exhibit strong time correlations. In this sense their collisions should show time correlation effects which should reflect in the friction term in equation (5.7). This term is dependent on the instantaneous velocity, but should include contributions from the velocity at previous times through a memory kernel (forming the generalized Langevin Dynamics), so that

$$m\frac{d\vec{v}}{dt} = -\int_{-\infty}^{t} dt'\ \gamma(t - t')v(t') + \vec{F}(t) \tag{5.9}$$

The random force is no longer uncorrelated and is constructed with correlations according to the fluctuation–dissipation theorem[1] ($\langle R(0)R(t)\rangle = \langle v^2\rangle\gamma(t)$). In Langevin dynamics, the Brownian particles are large and massive with respect to the thermal bath particles. The advantage is that no solvent particles are needed, making the use of longer time-steps and time scales up to second possible. The disadvantage is that the Langevin equation produces unphysical dynamics, as the Langevin thermostat is not Galilean-invariant, destroys the momentum conservation, and the hydrodynamic interactions are being screened.

---

[1] The response of a system in thermodynamic equilibrium to a small applied force is the same as its response to a spontaneous fluctuation [15].

5.2 Dissipative particle dynamics

Dissipative particle dynamics (DPD) is an off-lattice mesoscopic simulation scheme involving a set of particles moving in continuous space and discrete time [1, 16–18]. DPD is a coarse-graining of molecular mechanics and can simulate the dynamic and rheological properties of simple and complex fluids. The particles represent whole molecules or fluid regions moving coherently and reproducing the mesoscopic dynamics correctly. DPD is a stochastic method and the off-lattice version of lattice gas automata [19]. Within DPD, the dissipation–fluctuation theorem is enforced.

The original DPD model accounts for a collection of soft repelling, frictional and noisy balls. In principle, three are the types of forces which act between the dissipative particles: (a) **conservative** (F_{ij}^c), which are derived from a soft potential attempting to capture the effects of 'pressure' between different particles i and j, (b) **frictional**, which describe viscous resistance in real fluids, and (c) **stochastic**, which describe degrees of freedom eliminated in the coarse graining process, but responsible for the Brownian motion of the particles. The fundamental variables are the positions ($\vec{r}$) and velocities ($\vec{v}$) of the particles, obtained through

$$d\vec{r}_i = \vec{v}_i dt \qquad (5.10a)$$

$$m_i dv_i = \sum_{i \neq j} F_{ij}^c(r_{ij}) dt - \gamma \sum_{i \neq j} w(r_{ij})(\hat{e}_{ij}\vec{v}_{ij}) dt + \sigma \sum_{i \neq j} w^{1/2}(r_{ij})\hat{e}_{ij} dw_{ij} \qquad (5.10b)$$

where γ is the friction coefficient which controls the magnitude of the dissipative force, σ is the noise amplitude which defines the intensity of the stochastic force, and w a weight function, which provides the range of interaction for the dissipative particles. This is of a local nature as the particles interact only with their neighbors.

Equations (5.10) are translationally, rotationally, and Galilean invariant. The total momentum is preserved, as all three types of forces satisfy Newton's third law. All forces are pair-wise between particle center of mass, thus the linear/angular momentum is conserved. The softness of the potential prevents energy from diverging for long time steps. DPD captures the mass/momentum conservation responsible for the hydrodynamic behavior of a fluid at large scales and all stochastic and frictional forces are related through the fluctuation–dissipation theorem. DPD explores mesoscopic time scales and is a versatile scheme able to construct simple models for complex fluids. In equation (5.10) additional interactions increase the complexity of this scheme.

5.2.1 Stochastic rotation dynamics

Stochastic rotation dynamics (SRD) is a mesoscale scheme, which solves the linearized Navier–Stokes equations, based on local mass, momentum and energy conservation [4, 20, 21]. SRD is also known as multi-particles collision (MPC) [22, 23]. SRD simulations involve particles in a continuous space both in position and velocity, but evolves in discrete time steps. Similar to DPD, the particles are not water molecules, but coarse-grained particles which act as momentum and energy carriers. The SRD method consists of two basic steps: (i) a free streaming step and (ii) a collision step, which are performed at the same step. A system of N point particles of mass m with continuous coordinates $\vec{r}_i$ and velocities $\vec{v}_i$ is considered. The particles are initially assigned velocities according to a Gaussian distribution defined by the thermal energy of the system. Using other distributions would in the end also lead to velocities which follow a Gaussian distribution. Assuming a time step δt, the coordinates of a particle are updated during a streaming step according to:

$$\vec{r}_i(t + \delta t_{\text{MPC}}) = \vec{r}_i(t) + \vec{v}_i(t)\delta t_{\text{MPC}} \tag{5.11}$$

where $\vec{r}_i$ and $\vec{v}_i$ the position and velocity of particle i. The time step δt_{MPC} is stretched compared to atomistic MD and much longer time scales can be reached. The streaming step is followed by a collision, which considers the interactions between the particles. In order to perform the collision step involving multiple particles, a virtual grid is superimposed over the geometrical domain of the simulation box. Certain cells with a constant grid spacing a are assigned to the particles and the particle velocities within each cell are updated through

$$\vec{v}_i \rightarrow \vec{v}_{\text{CM}} + \hat{\mathbf{R}}\left(\vec{v}_i - \vec{v}_{\text{CM}}\right) \tag{5.12}$$

where $\vec{v}_{\text{CM}}$ is the center of mass velocity of the particles in the collision cell. The rotation matrix $\hat{\mathbf{R}}$ performs in two dimensions a rotation by an angle $+\alpha$ or $-\alpha$ each with a probability 1/2. In three dimensions, the rotation is the same for all particles in a cell but its direction is statistically independent between the cells. In order for the Galilean invariance to hold, a random shift in the collision cells is necessary. SRD is commonly used in the area of soft matter physics for polymer of colloid dynamics, as it can efficiently simulate fluids at low Reynolds numbers, but is not relevant to solid state physics.

5.2.2 Smoothed particle hydrodynamics

Smoothed particle hydrodynamics (SPH) is an alternative method for simulating fluids [6, 24]. It is a mesh free method in which the particle coordinates move with the simulated fluid. SPH uses a Langrangian description for the position and physical properties of the particles. Within SPH, the fluid is divided into discrete elements, which are the particles in this method. The relevant parameter is the smoothing length, h, which is the spatial distance of these particles. Over this length, a kernel function W (a Gaussian or a cubic spline) is used to smoothen the particle properties. For each particle, any physical property can be obtained through a summation of the relevant properties of all other particles in the range of the kernel. The contribution of each particle is weighted based on their distance from the particle of interest and the density. The latter also controls the resolution of SPH and the choice of the kernel function controls the particles which contribute to the properties of a certain particle. Any quantity A at any position r is then defined through the equation

$$A(\mathbf{r}) = \sum_j m_j \frac{A_j}{\rho_j} W\left(|\mathbf{r} - \mathbf{r}_j|, h\right), \tag{5.13}$$

where m_j is the mass of particle j, A_j is the value of the quantity A for the same particle, ρ_j is the density associated with that particle, r_j is the particle position, and $\mathbf{r}$ is the position in space at which the quantity A is evaluated.

An important adaptive feature of SPH is the possibility to assign a different smoothing length for each particle, which can also change in time. This can allow dense regions (with particles in short distances apart) to have a short smoothing length and less packed regions (with particles further apart) to have a larger smoothing length. Another way to change the adaptivity of SPH is by splitting the particles into 'daughter particles' with smaller smoothing lengths. SPH is a method widely used in astrophysics [25–27] or fluidics [28–30]. It has also been applied occasionally to solid state problems [23], like fracture of solids [31], fragmentation [32], shock wave propagation [33], etc.

5.3 Lattice methods

In contrast to continuum methods, lattice methods assume a grid with lattice points. Any equations (Navier–Stokes, continuity equation, etc) which need to be solved are solved on the discrete lattice points and not in the continuum space. By this, a lattice gas can be simulated, which maps discrete particles moving on a lattice from site to site and colliding. Again, a more mesoscopic behavior of the system is modeled and higher scales are reached.

5.3.1 Lattice Boltzmann method

The lattice Boltzmann method (LBM) is the basic grid-based computational method used to simulate fluids [3]. Within LBM the fluid is represented as a dilute gas in which particles move freely most of the time except when they collide. Such a dilute gas is described through the Boltzmann equation. The full Navier–Stokes equation can be derived from the Boltzmann equation, while different flavors of the LBM have been developed [34, 35]. Typically, all macroscopic variables are defined through particle distribution functions f_i that always reside on a lattice point. These denote probabilities of finding a particle at a certain grid point at a certain time. Accordingly, the local stress tensor $\Pi_{m,n}$, the macroscopic fluid density ρ, and the macroscopic velocity $\vec{u}$ can be obtained through summations over the distributions[2]:

$$\Pi_{m,n} = \sum_{i=0}^{\beta-1} \vec{e}_{i,m}\vec{e}_{i,n}\left(f_i - f_i^{\text{eq}}\right) \tag{5.14a}$$

$$\rho = \sum_{i=0}^{\beta-1} f_i \tag{5.14b}$$

---

[2] $\vec{e}_{ij}$ are directional vectors.

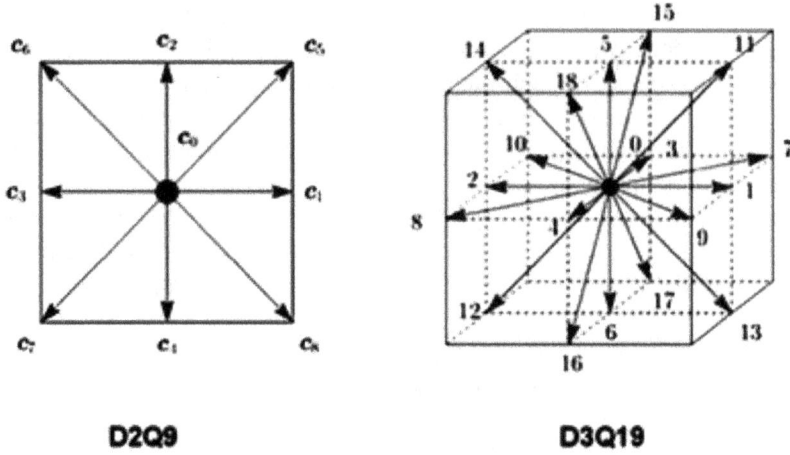

D2Q9　　　　　　　　　　　**D3Q19**

Figure 5.1. The two common LBM lattices: D2Q9 (left) and D3Q19 (right). The discrete velocities of each lattice are also shown by the arrows. Reproduced with permission from [36] and the Royal Society of Chemistry.

$$\vec{u} = \frac{1}{\rho}\sum_{i=0}^{\beta-1} f_i \vec{e}_i \qquad (5.14c)$$

The various lattices used with LBM are classified as $D_\alpha Q_\beta$, where α the space dimensionality and β the number of discrete velocities (with particles at rest within the momentum discretization). The most commonly used lattices are the D2Q9 in 2-D and the D3Q19 in 3-D sketched in figure 5.1.

The LBM algorithm involves streaming and collision for each time step and every mesh point. These two steps are handled separately. At each lattice point the populations are being updated according to:

$$f_i(\vec{x} + \vec{e}_i \Delta t, \, t + \Delta t) = f_i(\vec{x}, \, t) - \frac{\left| f_i(\vec{x}, \, t) - f_i^{\text{eq}}(\vec{x}, \, t) \right|}{\tau} \qquad (5.15)$$

with i spanning the momentum space and τ a relaxation parameter controlling the fluid viscosity. The fluid velocity is defined through $v = c_s^2(\tau - 1/2)$. Equation (5.15) is accurate for lattice points within the fluid domain, but not for any boundaries, as these need to be treated carefully. Regarding the two distinct steps in the LBM algorithm, during the streaming step, the distributions are translated to neighboring sites according to discrete velocity directions. The magnitude of the distribution functions remains unchanged, but moves to neighboring nodes according to one of the distinct directions defined by the specific lattice taken. During the collision step the distribution functions are re-distributed towards the local discretized Maxwell distribution function, so that the local mass and momentum are invariant. Viscous effects can be obtained through these collisions. The equilibrium distribution can be obtained from the local Maxwell–Boltzmann distribution. When the system is not in equilibrium, it will relax towards the equilibrium and the relaxation process can be very complicated. The relaxation is affected by the collision term in the Boltzmann equation. An important simplification in LBM is to approximate the collision

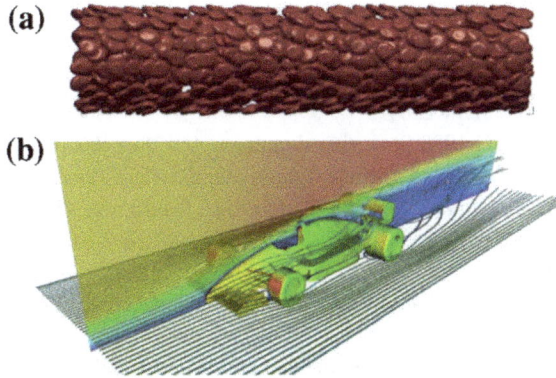

Figure 5.2. Examples of LBM modeling across different scales and disciplines. In (a) the distribution of deformable red blood cells (RBCs) in a pressure driven pipe flow are given using the LBM. Reproduced from [38] with permission from Elsevier. In (b) flow streamlines computed using LBM in a real-car geometry are shown. Reproduced from [39] with permission from the European Commission.

operator with the Bhatnagar–Gross–Krook (BGK) relaxation term [37] to define the collision term as

$$\left(\frac{\mathrm{d}f}{\mathrm{d}t}\right)_{\text{collision}} = -\frac{f - f^{\text{eq}}}{\tau} \tag{5.16}$$

The time evolution of a system in equilibrium with no collisions would be expressed through a simpler form like $f_i(\vec{r} + c_i \delta t, t + \delta t) = f_i(\vec{r}, t)$.

For treating the boundary conditions, different schemes are possible: (i) **periodic** for which the domain is folded along the direction of periodic boundary pair, (ii) **no-slip** for which the incoming distribution function at a wall node is reflected back to the original fluid nodes, (iii) **bounce-back**, which allows one to introduce obstacles into the fluid domain, as the incoming particles are populations with velocities pointing to an opposite direction. Overall, the advantage of LBM is that it can easily handle complex geometries. Compared to other methods it is quite flexible and can deal with large scale simulations (see figure 5.2) [40–42]. Multiphase/multicomponent LBM models have also been developed and can deal with mixtures and complex fluids [43]. LBM additionally can handle reactive flows [44], combustion (moving geometries) simulations [45], high Reynolds number flows [34], relativistic hydrodynamics [46], bubble simulations [47], colloids [48], electrophoretic motion [49] up to blood flows in arteries [50] or turbulent flows around macroscopic objects like a car [51] etc.

References

[1] Warren P B 1998 Dissipative particle dynamics *Curr. Opin. Colloid Interface Sci.* **3** 620–4
[2] Einstein A 1926 *Investigations on the Theory of the Brownian Movement* (London: Methuen)
[3] Succi S 2001 *The Lattice Boltzmann Equation: For Fluid Dynamics and Beyond* (Oxford: Oxford University Press)

[4] Ihle T and Kroll D M 2001 Stochastic rotation dynamics: a galilean-invariant mesoscopic model for fluid flow *Phys. Rev. E* **63** 020201

[5] Ebner C, Saam W F and Stroud D 1976 Density-functional theory of simple classical fluids. I. Surfaces *Phys. Rev. A* **14** 2264

[6] Brady J F and Bossis G 1988 Stokesian dynamics *Ann. Rev. Fluid Mech.* **20** 111–57

[7] Monaghan J J 2005 Smoothed particle hydrodynamics *Rep. Prog. Phys.* **68** 1703

[8] Thijssen J 2007 *Computational Physics* (Cambridge: Cambridge University Press)

[9] Allen M P and Tildesley D J 1989 *Computer Simulation of Liquids* (Oxford: Oxford University Press)

[10] Steinhauser M O 2008 *Computational Multiscale Modeling of Fluids and Solids* (Berlin: Springer)

[11] Batchelor G K 2000 *An Introduction to Fluid Dynamics* (Cambridge: Cambridge university press)

[12] Versteeg H K and Malalasekera W 2007 *An Introduction to Computational Fluid Dynamics: The Finite Volume Method* (Harlow: Pearson)

[13] Boon J P, Yip S, Burgers J M, Van de Hulst H C, Chandrasekhar S, Chen F F, De Loore C B, Frenkel D and Smit B 1988 *An Introduction to Fluid Mechanics* (Cambridge: Cambridge University Press)

[14] Langevin P 1908 Sur la théorie du mouvement brownien *CR Acad. Sci. Paris* **146** 530

[15] Chandler D 1987 *Introduction to Modern Statistical Mechanics* (Oxford: Oxford University Press)

[16] Espanol P 2005 Dissipative particle dynamics *Handbook of Materials Modeling* (Berlin: Springer) pp 2503–12

[17] Pivkin I V, Caswell B, Karniadakis G E and Lipkowitz K 2011 Dissipative particle dynamics *Rev. Comput. Chem.* **27** 85–110

[18] Groot R D *et al* 1997 Dissipative particle dynamics: Bridging the gap between atomistic and mesoscopic simulation *J. Chem. Phys.* **107** 4423

[19] Wolf-Gladrow D A 2000 *Lattice-Gas Cellular Automata and Lattice Boltzmann Models: An Introduction* (Berlin: Springer)

[20] Petersen M K *et al* 2010 Mesoscale hydrodynamics via stochastic rotation dynamics: comparison with lennard-jones fluid *J. Chem. Phys.* **132** 174106

[21] Pooley C M and Yeomans J M 2005 Kinetic theory derivation of the transport coefficients of stochastic rotation dynamics *J. Phys. Chem. B* **109** 6505–13

[22] Malevanets A and Kapral R 1999 Mesoscopic model for solvent dynamics *J. Chem. Phys.* **110** 8605–13

[23] Gompper G, Ihle T, Kroll D M and Winkler R G 2009 Multi-particle collision dynamics: a particle-based mesoscale simulation approach to the hydrodynamics of complex fluids *Advanced Computer Simulation Approaches for Soft Matter Sciences III* (Berlin: Springer) pp 1–87

[24] Stellingwerf R F 1991 Smooth particle hydrodynamics *Advances in the Free-Lagrange Method Including Contributions on Adaptive Gridding and the Smooth Particle Hydrodynamics Method* (Berlin: Springer) pp 239–47

[25] Dolag K, Bartelmann M and Lesch H 1999 SPH simulations of magnetic fields in galaxy clusters arXiv preprint astro-ph/9906329

[26] Berczik P 1999 Chemo-dynamical SPH code for evolution of star forming disk galaxies Arxiv preprint astro-ph/9907375

[27] Owen J M, Villumsen J V, Shapiro P R and Martel H 1998 Adaptive smoothed particle hydrodynamics: Methodology. ii *Astrophys. J. Suppl. Ser.* **116** 155

[28] Viccione G, Bovolin V and Carratelli E P 2008 Defining and optimizing algorithms for neighbouring particle identification in sph fluid simulations *Int. J. Numer. Methods Fluids* **58** 625–38

[29] Monaghan J J 1994 Simulating free surface flows with sph *J. Comput. Phys.* **110** 399–406

[30] Yan H, Wang Z, He J, Chen X, Wang C and Peng Q 2009 Real-time fluid simulation with adaptive sph *Comput. Animat. Virtual Worlds* **20** 417–26

[31] Hoover W G and Hoover C G 2001 Spam-based recipes for continuum simulations *Comput. Sci. Eng.* **3** 78–85

[32] Rabczuk T and Eibl J 2003 Simulation of high velocity concrete fragmentation using SPH/MPSPH *Int. J. Numer. Methods Eng.* **56** 1421–44

[33] Herreros M I and Mabssout M 2011 A two-steps time discretization scheme using the SPH method for shock wave propagation *Comput. Methods Appl. Mech. Eng.* **200** 1833–45

[34] Boghosian B M, Yepez J, Coveney P V and Wager A 2001 Entropic lattice Boltzmann methods *Proc. R. Soc. A* **457** 717–66

[35] Ladd A J C and Verberg R 2001 Lattice-Boltzmann simulations of particle-fluid suspensions *J. Stat. Phys.* **104** 1191–251

[36] Kondaraju S, Farhat H and Lee J S 2012 Study of aggregational characteristics of emulsions on their rheological properties using the lattice Boltzmann approach *Soft Matter* **8** 1374–84

[37] Bhatnagar P L, Gross E P and Krook M 1954 A model for collision processes in gases. i. small amplitude processes in charged and neutral one-component systems *Phys. Rev.* **94** 511

[38] Mills Z G, Mao W and Alexeev A 2013 Mesoscale modeling: solving complex flows in biology and biotechnology *Trends Biotechnol.* **31** 426–34

[39] Test D *et al* 2011 *STREP FP7-248032: Multiscale Spatiotemporal Visualisation* (Brussels: European Commission)

[40] Benzi R, Succi S and Vergassola M 1992 The lattice Boltzmann equation: theory and applications *Phys. Rep.* **222** 145–97

[41] Harting J, Chin J, Venturoli M and Coveney P V 2005 Large-scale lattice Boltzmann simulations of complex fluids: advances through the advent of computational grids *Phil. Trans. R. Soc.* **363** 1895–915

[42] Zhang J 2011 Lattice Boltzmann method for microfluidics: models and applications *Microfluid. Nanofluid.* **10** 1–28

[43] Shan X and Chen H 1993 Lattice Boltzmann model for simulating flows with multiple phases and components *Phys. Rev. E* **47** 1815

[44] Di Rienzo A F, Asinari P, Chiavazzo E, Prasianakis N I and Mantzaras J 2012 Lattice Boltzmann model for reactive flow simulations *EPL* **98** 34001

[45] Chiavazzo E, Karlin I V, Gorban A N and Boulouchos K 2009 Combustion simulation via lattice Boltzmann and reduced chemical kinetics *J. Stat. Mech.* **2009** P06013

[46] Mendoza M, Boghosian B M, Herrmann H J and Succi S 2010 Derivation of the lattice Boltzmann model for relativistic hydrodynamics *Phys. Rev. D* **82** 105008

[47] Sankaranarayanan K, Kevrekidis I G, Sundaresan S, Lu J and Tryggvason G 2003 A comparative study of lattice Boltzmann and front-tracking finite-difference methods for bubble simulations *Int. J. Multiph. Flow* **29** 109–16

[48] Ladd A J C 1993 Short-time motion of colloidal particles: numerical simulation via a fluctuating lattice-Boltzmann equation *Phys. Rev. Lett.* **70** 1339

[49] Lobaskin V, Dünweg B, Medebach M, Palberg T and Holm C 2007 Electrophoresis of colloidal dispersions in the low-salt regime *Phys. Rev. Lett.* **98** 176105

[50] Artoli A M, Hoekstra A G and Sloot P M A 2003 Simulation of a systolic cycle in a realistic artery with the lattice Boltzmann BGK method *Int. J. Mod. Phys.* B **17** 95–8

[51] Chen H, Kandasamy S, Orszag S, Shock R, Succi S and Yakhot V 2003 Extended Boltzmann kinetic equation for turbulent flows *Science* **301** 633–6

Chapter 6

The Monte Carlo method

The computational methods reviewed up to this chapter are all deterministic. On the other hand, a stochastc determination of the properties of a system is also possible. The most common scheme for this is the Monte Carlo (MC) method, first introduced in the 1950s [1–4]. The stochastic nature of this scheme, applicable to both classical and quantum-mechanical problems, is based on the use of random numbers generated during a computer simulation [5], which more or less determine the regions of the phase space being sampled. Typical examples of using the MC method in physics involve and go beyond the calculation of thermodynamic quantities in statistical physics (e.g. Ising model magnetization [6, 7]), multiple integrals, reaction-diffusion problems (in geophysics or chemical engineering), polymer, protein conformations, particle accelerators, surface phenomena, etc. MC allows the simulation of many-particle systems by introducing artificial dynamics based on random numbers. The artificial dynamics in MC prevents its use for determining dynamic physical properties, but for **static** properties it is very fast and efficient. The main types of MC simulations are: (i) **direct MC** in which random numbers are used to model the effect of complicated properties, the details of which are not crucial (e.g. modeling traffic and the behavior of cars is determined by random numbers). (ii) **MC integration** can be used for the calculation of integrals using random numbers and is efficient for integration over a high-dimensional space. (iii) **Metropolis MC**, in which the system undergoes transitions from one state to another in a random, memory-less process. It can be shown, that the error in the MC integration decreases as $N^{-1/2}$. The computational cost is relative to the total number of sub-intervals n or the total number of points N. Accordingly, a MC integration is more efficient for $D > 2a$. MC can be easily applied to simple mathematical problems, such as the calculation of π, integrals, sampling of the mean, etc [8, 9]. A good understanding of the MC method can be acquired through its typical example, the Ising model [7, 10, 11]. In the following, focus will be given on how the method is applied to physical problems.

doi:10.1088/978-1-6817-4417-9ch6

6.1 Random numbers

Before moving into the main details of MC methods, a short overview of random numbers [5] is attempted. These are tightly connected with the stochastic nature of MC. Random numbers are usually distributed uniformly over the interval [0, 1]. Each number within this interval has the same probability of occurrence and cannot be predicted. The random numbers are computer generated and only values on a dense discrete set are possible due to the finite number of bits used in their representation. An alternative would be to generate nonuniform random numbers, in which case some numbers have a higher occurrence probability. In principle, computer generated random numbers are not always truly random, as they are often generated from previous ones through a mathematical formula. Nevertheless, pseudo-random numbers are in practice sufficient as they are a sequence of numbers difficult to distinguish from sequences of pure random numbers. One of the main criteria for the quality of random numbers involves the absence of correlations within their distributions.

A random number generator (RNG) is a physical or computational device designed to generate a sequence of random numbers. A pseudo-RNG is an algorithm, which can automatically create long runs of numbers with good random properties. Pseudo-RNG are always represented by the finite number of bits. The string of values generated by such algorithms is determined by a fixed number, the **seed**. Random numbers of a good quality (in terms of randomness) should have a long period before repeat and should not be explicitly dependent on computer specific handling or overflow in integer multiplication. They should also be portable and not show a strong correlation. One should always have in mind, that an RNG can be good for one simulation method, but not for another. Further details can be found in the literature [12–15].

6.2 Classical Monte Carlo

In principle, the MC method involves how the sampling of the conformational space of a system is being performed in order to get ensemble averages and obtain static properties. For this, different ways of sampling are possible, which are related to the random motion of the system particles and the acceptance (or not) of this movement. In the following, the different sampling and acceptance techniques typically used in MC simulations in physics will be briefly reviewed.

6.2.1 Sampling of the phase space

Simple sampling samples homogeneously, that is, unweighted averages of a function $f(x)$ are determined by evaluating $f(x)$ at a large number of random x values, which are homogeneously distributed over an interval [a, b] [16]. In this way, the whole space within [a, b] is sampled. This is often not optimal for two main reasons: it becomes computationally too expensive to sample over the whole phase space and it is not relevant when a specific region of the phase space is very important and needs to be sampled more efficiently.

Importance sampling assures that certain values of the input random variables in a simulation have more impact on the parameter being estimated than others [17–19]. In this way, a certain part of the phase space of the problem is being sampled. The main characteristic of importance sampling is to choose a distribution, which encourages sampling of these regions of interest. For this, biased distributions come into play and need to be chosen correctly. Unreliable results and longer simulation times are the result of non-optimum biased distributions.

For x the sampling value and $f(x)$ the relevant probability function, $g(x)$ would be the biased distribution function with $f(x)/g(x)$ the so-called likelihood ratio which links to the weights taken for each distribution. In this respect, a higher weight will be assigned to the distributions that need to be sampled more frequently. The results from the simulations are then corrected for the use of the biased distribution, so that the next sampling step is unbiased. Other types of algorithms like the adaptive MC schemes [20–22] also aim at regions where $f(x)$ contributes significantly. These methods locate these regions by probing the function at random points and require no *a priori* knowledge of $f(x)$ as in importance sampling.

6.2.2 Markov chains

Importance sampling can be realized using the concept of Markov chains (or Markov process) [39]. In a Markov chain, each state in the system (or each part of the phase space) X_{i+1} is constructed from a previous state X_i through a suitable transition probability $W(x_i \rightarrow X_{i+1})$: For uncorrelated chains the probability of occurrence of a sequence of N objects $X_1, ..., X_N$ is statistically uncorrelated and $P_N(X_1, ..., X_N) = P_1(X_1) ... P_N(X_N)$. In a Markov chain, a non-zero probability is assigned to the system remaining in the same state, that is $W(X \rightarrow X') > 0$ $\forall \{X\}, \{X'\}$. The Markov state must also be some state when starting from $\{X\}$ and the normalization condition $\sum_{X'} W(X \rightarrow X') = 1, \forall_{\{X\}}$ should hold. The Markov chains are completely specified by joint probabilities and successive transitions are independent. The Markov chain is defined through transition probabilities $P(X \rightarrow X')$ for having state X' succeed state X. The probability of having a sequence of objects X_i would be:

$$P_N(X_1, X_2, ..., X_N) = P_1(X_1)W(X_1 \rightarrow X_2)W(X_2 \rightarrow X_3)...W(X_{N-1} \rightarrow X_N) \quad (6.1)$$

Markov processes assure also that equilibrium distributions (P^{eq}) do not get destroyed once reached, that is

$$\sum_X W(X \rightarrow X')P^{eq}(X) = P^{eq}(X') \quad (6.2)$$

In statistical physics the equilibrium distribution is the Boltzmann distribution:

$$P^{eq}(X) = \frac{1}{Z}\exp\left[-\beta \mathcal{H}(X)\right] \quad (6.3)$$

where $\mathcal{H}$ is the Hamiltonian of the system, β the thermal energy, and Z the relevant partition function. Note, that ergodicity is essential in order to reach the desired P^{eq}.

In this respect, each state should be **irreducible** and **aperiodic**. Irreducibility means that every configuration included in the ensemble should be accessible from every other configuration by a finite number of steps. The transitional probability should be non-zero, otherwise the configuration is absorbing. Aperiodicity is related to the fact that after visiting a configuration, it should be possible to return to the same configuration, except after $t = nk$ steps with $n = 1, 2, \ldots$ and k fixed.

6.2.3 Detailed balance

For reversible Markov chains, the product of transition rates over any closed loop of states must be the same in both directions. $P(X, t)$ is again the probability of occurrence of a configuration X given at 'time' (or Markov step) t [1]. For an ergodic chain, $P(X, t)$ becomes independent of t for large t. The rate of change in $P(X, t)$ from one step to another arises from two processes: (i) moving from X at t to some X' at $t + dt$ (decreasing $P(X)$) and (ii) moving from some X' at t to X at $t + dt$ (increasing $P(X)$). The master equation for the change in the occurrence probability giving the direction of a move would then be

$$\frac{\mathrm{d}}{\mathrm{d}t}P(X, t) = -\sum_{X'}W(X \to X')P(X, t) + \sum_{X'}W(X' \to X)P(X', t) \quad (6.4)$$

In this equation, $W(X \to X') = \omega_{xx'}A_{xx'}$, where $\omega_{xx'}$ is the trial step probability and $A_{xx'}$ the acceptance probability of a move. The matrix ω is symmetric ($\omega_{xx'} = \omega_{x'x}$) and the normalization condition $\sum_{x'}\omega_{xx'} = 1$ holds, as well as $0 \leqslant \omega_{xx'} \leqslant 1$. $\forall x, x'$ the acceptance probability is bound, so that $0 < A_{xx'} < 1$. In order to find a stationary solution, which is invariant in time, $\frac{\mathrm{d}P(X, t)}{\mathrm{d}t} = 0$. The detailed balance condition is then

$$\frac{A_{xx'}}{A_{x'x}} = \frac{P(X')}{P(X)} \quad (6.5)$$

For $P(X)$ being the Boltzmann distribution,

$$\frac{W(X \to X')}{W(X' \to X)} = \frac{P(X')}{P(X)} = \frac{e^{-\beta\mathcal{H}(X')}}{e^{-\beta\mathcal{H}(X)}} = e^{-\beta\Delta\mathcal{H}} \quad (6.6)$$

and the detailed balance implies that the ratio of transition probabilities for a move and its inverse move can only depend on the respective energy change. A good choice for $W(X \to X')$ is proposed in the Metropolis algorithm.

6.2.4 Umbrella sampling

A method which attempts to improve the sampling of a system is umbrella sampling [23]. A poor sampling is often related to an existing energy barrier separating two regions of the configuration space. Umbrella sampling attempts to overcome this

---

[1] t is not a real time, but rather corresponds to the number of MC steps.

barrier by means of bias potentials. The bias potentials along a reaction coordinate (in one or more dimensions) can drive the system from one thermodynamic state to another. This reaction pathway is divided into windows, as sketched in figure 6.1. At each of these windows a simulation (MD or MC) is performed. In each window, the sampling can be improved by other schemes by including simulations at higher temperatures or through exchange between successive windows like in replica exchange methods [24]. Although, any function could be used as the bias potential, harmonic ones are used for simplicity. These bias potentials constrain the system in each window. From the sampling along the reaction coordinate, the free energy difference in each window can be calculated and corrected for the biasing potential. The adaptive-bias umbrella sampling can adapt the bias potential in a way to produce an even distribution between the states. In this case, the whole range can be spawned by one window and the bias voltage results directly in the free energy difference [25].

In order to make the umbrella sampling method more intuitive, take ξ to be the reaction coordinate and w_i the bias potential for window i, which depends only on the reaction coordinate. If r is the position coordinate, while 'b' and 'u' denotes biased and unbiased, then $E^b(r) = E^u(r) + w_i(\xi)$. The probability distribution of the system along the reaction coordinate, would be expressed as

$$P_i^{(u,b)} = \frac{\int \exp\left[-\beta E^{(u,b)}(r)\right]\delta\left[\xi'(r) - \xi\right]d^N r}{\int \exp\left[-\beta E^{u,b}(r)\right]d^N r} \quad (6.7)$$

P_i^b is obtained from a biased simulation and the biased potential is known analytically. Note, that ergodicity is assumed.

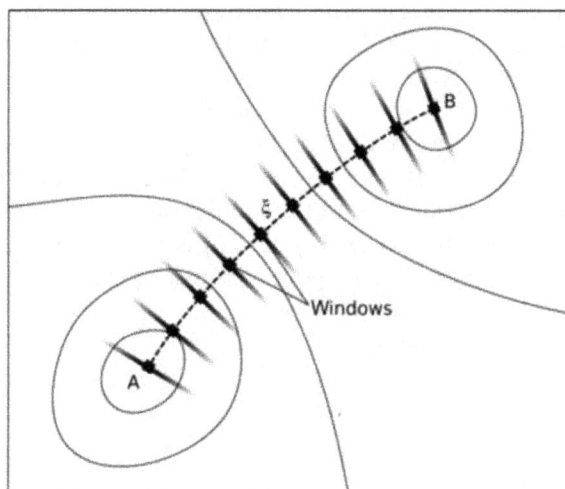

Figure 6.1. Separation of the reaction coordinate (dashed line) between two states (here represented by two minima on the potential energy surface, A and B) into distinct windows. The system is mainly sampled perpendicular to the reaction coordinate in each window. Reproduced from [25] with permission from Wiley.

6.2.5 Metropolis algorithm

One of the widely used acceptance algorithms in MC is the Metropolis algorithm [1], which is conceptually simple and easy to be implemented and is based on detailed balance. A general flow of the algorithm is presented here:

- Start from a state X with $P(X) > 0$.
- Propose a new state X' with a probability given by $\omega_{XX'}$.
- Compare $P(X)$ with $P(X')$ and calculate the acceptance probability:

$$
A_{XX'} = \begin{cases} 1, & \text{if } P(X') > P(X) \\ \dfrac{P(X')}{P(X)}, & \text{if } P(X') < P(X) \end{cases} \tag{6.8}
$$

- Generate a random number r uniformly distributed in $[0, 1]$. If $r < A_{XX'}$, the new state X' is accepted, with a probability $A_{XX'}$. If $r \geqslant A_{XX'}$, the new state X' is rejected with a probability $1 - A_{XX'}$. When the new state X' is accepted, it replaces the old state X. When the new state X' is not accepted, the system remains in its old state X.
- Update and start again with a new state X'.

Note, that the new state X' is always accepted if $P(X') > P(X)$. A very natural and common choice of the transition probability according to the Metropolis algorithm for simulating physical systems is the Boltzmann distribution:

$$
W(X \to X') = \min[1, \exp(-\beta \Delta \mathcal{H})] \tag{6.9}
$$

where $\Delta \mathcal{H} = \mathcal{H}(X') - \mathcal{H}(X)$ is the Hamiltonian between the old and the new state.

Within the Metropolis algorithm the configurations are generated in a Markov chain and include correlations. The theory of Markov chains guarantees that at long times the distribution $P(X)$ is indeed invariant. The problem is that these times might be longer than the available computing time. The total number of statistically independent configurations is given by the total number of steps divided by the correlation 'time'. This is measured in MC steps, which are the total number of trials (not the successful trials only). For n the total MC steps and n_0 the necessary equilibration steps at the initial stage of the simulation, a sequence of configurations $\{X\}$ with a statistical distribution $\exp[-\beta \mathcal{H}(X))]$ is generated with the Metropolis algorithm leading to the ensemble average of a physical quantity A, expressed through

$$
\langle A \rangle = \frac{1}{n - n_0} \sum_{\nu > n_0}^{n} A_\nu. \tag{6.10}
$$

6.2.6 More on acceptance probabilities

The Boltzmann distribution is indeed a natural choice of the transition probability, but other choices could also be efficient. Detailed balance is fulfilled by the

Metropolis algorithm, but is valid also within other types of acceptance algorithms, such as the ones given here:

- Barker algorithm [26]:

$$W(X \to X') = \omega_{XX'} \frac{P(X')}{P(X) + P(X')}, \qquad X \neq X' \qquad (6.11a)$$

$$W(X \to X) = 1 - \sum_{X \neq X'} W(X \to X') \qquad (6.11b)$$

where $\omega_{XX'}$ satisfies the same criteria as in the Metropolis algorithm.
- Generalized Metropolis (Metropolis–Hastings) algorithm [27]: For some pairs X, X' of configurations $\omega_{XX'} \neq \omega_{X'X}$ is taken. In those cases the acceptance criteria for $X \to X'$ must be replaced by $A_{XX'} = \min(1, q_{XX'})$ with $q_{XX'} = \frac{\omega_{X'X} P(X')}{\omega_{XX'} P(X)}$ which satisfies detailed balance.
- Glauber algorithm [28]: The acceptance probability of a local move is defined as

$$W(X \to X') = \frac{1}{2}\left[1 - \tanh\left(\frac{\beta \Delta \mathcal{H}}{2}\right)\right] \qquad (6.12)$$

where again β is the inverse temperature and $\Delta \mathcal{H}$ is the difference in energy for moving from state X to state X'. The Glauber algorithm in the case of a Markov chain (as in the Ising model [6, 7]) can be calculated analytically in one dimension, but no exact solutions exist for higher dimensions.
- Heat bath algorithm [29]: This algorithm assumes that the trial step involves one or a few degrees of freedom and the remaining ones are kept fixed. The degrees of freedom that change are x, while the remainder are $X - x$. The acceptance probability is defined as:

$$P(x) \propto \exp\left[-\beta \mathcal{H}(x/X - x)\right] \qquad (6.13)$$

where $\mathcal{H}(x/X - x)$ is the Hamiltonian as a function of x with $X - x$ fixed. This algorithm also satisfies detailed balance and is equivalent to applying an infinite number of Metropolis steps to x successively, with fixed $X - x$. The implementation of this algorithm is often difficult and becomes easier for lattice models and only discrete degrees of freedom with a few allowed states.

6.2.7 An example: solving the Ising model with MC

In classical MC, the calculation of the partition function is obtained through sampling and is important for calculating the average properties, The partition function can be estimated either through a direct calculation or by efficiently sampling the phase space through N representative configurations using an MC scheme with a specific probability distribution. A direct calculation can only be done for small lattices (e.g. a 2-D Ising lattice model) as a too long computational time will be needed otherwise. Typically the simulations are performed through a scheme of MC sampling, which for the Ising model will be sketched as follows [30, 31]:

- Generate a random configuration of spins on a lattice and apply periodic boundary conditions. Calculate the initial energy $E_0 = -J\sum_{\langle ij\rangle}\sigma_i^z\sigma_j^z - B\sum_i\sigma_i^z$, where J, B, and σ_j^z are the coupling constant, magnetic field and spin i, respectively.
- Randomly select a lattice site i with spin σ_i^z and flip it: $\sigma_i^z \to -\sigma_i^z$.
- Calculate the energy change due to the flip: $\Delta\mathcal{H} = \mathcal{H}(\mu) - \mathcal{H}(\nu)$ with $\mu = \sigma_i^{z'}$ the new state after flipping and $\nu = \sigma_i^z$ the old state.
- Calculate the transition probability for the flip. For a Metropolis algorithm:
  - for $\Delta\mathcal{H} \leqslant 0$, always accept the flip,
  - for $\Delta\mathcal{H} > 0$, evaluate $P = e^{-\beta\Delta\mathcal{H}}$ and draw a random number $r \in (0, 1)$ uniformly distributed in $(0, 1)$. If $P > r$, the flip is accepted and the energy of the new state is the initial energy $E_0 = \mathcal{H}(\mu)$ for the next step. For $P \leqslant r$, the flip is rejected.

The algorithm can be performed at a chosen fixed temperature or include all steps from step one and on in a loop over temperatures. There is a unique probability distribution of the states at each temperature. Performing the loop, evolves the Markov chain until a stationary distribution has been reached.

6.3 Quantum Monte Carlo (QMC)

MC techniques can also be applied to quantum mechanical systems in order to calculate the properties of a collection of interacting quantum mechanical particles. In quantum Monte Carlo (QMC) [32–35], the many-body wavefunction is modeled directly, typically using Hartree–Fock as a starting point. QMC schemes exist which are of a static and dynamic nature, that is time independent or time-dependent [33–37]. Different algorithms [8] have been developed and applied in QMC, such as:

- Variational MC [38]: it optimizes the expectation value by adjusting a trial wavefunction in a variational approach.
- Diffusion MC [41, 42]: it uses the similarity of the Schrödinger and diffusion equations to calculate the properties of a quantum system.
- Path-integral MC [41] in which the quantum mechanical problem is mapped on a classical one.
- Reptation MC [42]: zero-temperature and is related to path integral MC.
- Stochastic Green's function [43]: simulates any complicated lattice Hamiltonian with no sign problem and is mainly designed for bosons.
- Auxiliary field MC [44] is used for lattice problems.

Example: In order to elucidate the way a QMC method is applied, the variational MC is taken as an example. The goal is to find the ground state and the first few excited states of a quantum Hamiltonian. This would necessitate the calculation of high-dimensional integrals. One needs first to recall the variational method, which is constructed by the following steps:

i. Construction of the trial $\psi_a(r)$ depending on S variational parameters $a = (a_1, a_2, ...,a_S)$

ii. Evaluation of the expectation value

$$\langle E \rangle = \frac{\langle \psi_a|H|\psi_a \rangle}{\langle \psi_a|\psi_a \rangle}$$

iii. Variation of a with respect to the minimization algorithm and return to step (i).

The loop stops when the minimum energy is reached according to some criteria. QMC is performed at step (ii).

In realistic many-body systems, the wavefunctions have small values in large parts of the configuration space. This problem can be overcome through the Metropolis algorithm in which a collection of random walkers are pushed towards regions where the wavefunctions have appreciable values. The term $H\psi_t$ needs to be evaluated for any trial ψ_t and the local energy is defined as

$$E_L(r) = \frac{H\psi_t(r)}{\psi_t(r)} \tag{6.14}$$

The energy $E_L(r)$ depends on particle positions and is constant if ψ_t is exact. $E_L(r)$ varies less the closer ψ_t approaches the exact wavefunction. The expectation value is then

$$\langle E \rangle = \frac{\int dr\ \psi_t^2(r)E_L(r)}{\int dr\ \psi_t^2(r)} \tag{6.15}$$

It is possible to construct a Metropolis walk with a stationary distribution $\rho(r) = \frac{\psi_t^2(r)}{\int dr'\ \psi_t^2(r')}$ to solve the problem, similarly as in the classical case (see Ising model).

References

[1] Metropolis N, Rosenbluth A W, Rosenbluth M N, Teller A H and Teller E 1953 Equation of state calculations by fast computing machines *J. Chem. Phys.* **21** 1087–92

[2] Binder K 1995 *Monte Carlo and Molecular Dynamics Simulations in Polymer Science* vol 95 (New York: Oxford University Press)

[3] Binder K and Heermann D 2010 *Monte Carlo Simulation in Statistical Physics: An Introduction* (Berlin: Springer)

[4] Landau D P and Binder K 2014 *A Guide to Monte Carlo Simulations in Statistical Physics* (Cambridge: Cambridge University Press)

[5] Hamming R 1973 *Numerical Methods for Scientists and Engineers* (New York: Dover)

[6] Ising E 1925 A contribution to the theory of ferromagnetism *Z. Phys.* **31** 253–8

[7] Ising E 1925 Beitrag zur theorie des ferromagnetismus *Zeitschrift Für Physik A Hadrons and Nuclei* **31** 253–8

[8] Thijssen J 2007 *Computational Physics* (Cambridge: Cambridge University Press)

[9] Liu J S 2008 *Monte Carlo strategies in scientific computing* (Berlin: Springer)

[10] Lenz W 1920 Beitrag zum verständnis der magnetischen erscheinungen in festen körpern *Z. Phys.* **21** 613–5

[11] Onsager L 1944 Crystal statistics. i. a two-dimensional model with an order–disorder transition *Phys. Rev.* **65** 117

[12] Gentle J E 2006 *Random Number Generation and Monte Carlo Methods* (Berlin: Springer)

[13] L'ecuyer P 1988 Efficient and portable combined random number generators *Commun. ACM* **31** 742–51

[14] L'Ecuyer P 2012 Random number generation *In Handbook of Computational Statistics* (Berlin: Springer) pp 35–71

[15] Hull T E and Dobell A R 1962 Random number generators *SIAM Rev.* **4** 230–54

[16] Tillé Y 2011 *Sampling Algorithms* (Berlin: Springer)

[17] Srinivasan R 2013 *Importance Sampling: Applications in Communications and Detection* (Berlin: Springer)

[18] Press W H 2007 *Numerical Recipes 3rd edition: The Art of Scientific Computing* (Cambridge: Cambridge University Press)

[19] Veach E and Guibas L J 1995 Optimally combining sampling techniques for Monte Carlo rendering *Proc. of the 22nd Annual Conference on Computer Graphics and Interactive Techniques ACM* pp 419–28

[20] Lapeyre B and Lelong J 2011 A framework for adaptive Monte Carlo procedures *Monte Carlo. Methods Appl.* **17** 77–98

[21] Dimov I T and McKee S 2008 *Monte Carlo Methods for Applied Scientists* vol 42 (Singapore: World Scientific)

[22] Martino L, Elvira V, Luengo D and Corander J 2015 An adaptive population importance sampler: learning from uncertainty *IEEE Trans. Signal Process.* **63** 4422–37

[23] Torrie G M and Valleau J P 1977 Nonphysical sampling distributions in Monte Carlo free-energy estimation: Umbrella sampling *J. Comput. Phys.* **23** 187–99

[24] Swendsen R H and Wang J- S 1986 Replica Monte Carlo simulation of spin-glasses *Phys. Rev. Lett.* **57** 2607

[25] Kästner J 2011 Umbrella sampling *WIREs: Comput Mol Sci.* **1** 932–42

[26] Barker A A 1965 Monte Carlo calculations of the radial distribution functions for a proton–electron plasma *Aust. J. Phys.* **18** 119–34

[27] Hastings W K 1970 Monte Carlo sampling methods using Markov chains and their applications *Biometrika* **57** 97–109

[28] Glauber J J 1963 Time-dependent statistics of the Ising model *J. Math. Phys.* **4** 294–307

[29] Henkel M, Pleimling M and Sanctuary R 2007 *Ageing and The Glass Transition* vol 716 (Berlin: Springer)

[30] Binder K 2001 Ising model *Hazewinkel, Michiel, Encyclopedia of Mathematics* (Berlin: Springer) p 1

[31] McCoy B M and Wu T T 1973 *The Two-Dimensional Ising Model* (Cambridge, MA: Harvard University Press)

[32] Ceperley D and Alder B 1986 Quantum Monte Carlo *Science* **231** 555–60

[33] Suzuki M (ed) 1993 Quantum Monte Carlo methods in condensed matter physics (Singapore: World Scientific)

[34] Foulkes W M C *et al* 2001 Quantum Monte Carlo simulations of solids *Rev. Mod. Phys.* **73** 33

[35] Hammond B L, Lester W A Jr and Reynolds P J 1994 Monte Carlo methods in ab initio quantum chemistry Vol. 1 (Singapore: World Scientific)

[36] Ceperley D M and Bernu B 1988 The calculation of excited state properties with quantum Monte Carlo *J. Chem. Phys.* **89** 6316–28

[37] Umrigar C J, Nightingale M P and Runge K J 1993 A diffusion Monte Carlo algorithm with very small time-step errors *J. Chem. Phys.* **99** 2865–90

[38] Fahy S, Wang X W and Steven G L 1988 Variational quantum Monte Carlo nonlocal pseudopotential approach to solids: cohesive and structural properties of diamond *Phys. Rev. letters* **61** 1631

[39] Gilks W R 2005 *Markov Chain Monte Carlo* (New York: Wiley)

[40] Rapaport D C 2004 *The Art of Molecular Dynamics Simulation* (Cambridge: Cambridge University Press)

[41] Barker J A 1979 A quantum-statistical Monte Carlo method; path integrals with boundary conditions *J. Chem. Phys.* **70** 2914–8

[42] Baroni S and Moroni S 1999 Reptation quantum Monte Carlo: a method for unbiased ground-state averages and imaginary-time correlations *Phys. Rev. letters* **82** 4745

[43] Rousseau V G 2008 Stochastic Green function algorithm *Phys. Rev. E* **77** 056705

[44] Honma M, Mizusaki T and Otsuka T 1995 Diagonalization of Hamiltonians for many-body systems by auxiliary field quantum Monte Carlo technique *Phys. Rev. letters* **75** 1284

Chapter 7

Multiscale, hybrid, and coarse-grained methods

7.1 Coarse-graining

All chemical and atomistic details of a physical system are often not very relevant in defining the property of interest. In these cases, including all the fine degrees of freedom of the system significantly increases the computational effort without adding any important features in the results. It is then good practice to reduce the number of the degrees of freedom of the system in order to decrease the computational cost [1–4]. The reduction of degrees of freedom needs to be done in a way that the information relevant to the desired properties is not lost. This is typically done by modeling coarse-grained particles in the form of spherical beads. Depending on the coarse-graining (CG) level, one bead can map a number (ranging from a few to a few thousands) of particles (atoms, molecules, residues, etc).

A sketch of a CG is given in figure 7.1. The beads in the coarse-grained simulations include implicitly the information regarding the particles, which have been coarse-grained. The CG simulations are being carried out with the coarse-grained particles (beads) together with the relevant CG potentials. In this respect, additionally to the CG, potentials modeling the interactions of the coarse-grained particles need to be carefully chosen or developed. In the end, the results from the simulations can be mapped back to the real particles and the finer scales. Resort to CG allows an alternative and efficient modeling of a wide range of problems related to molecular and biomolecular systems [5, 6] and colloidal suspensions [7] up to red blood cells [8]. It also allows the modeling of systems not easily tackled with one of the available all-atom methods. In the end, CG depends on the systems, the methods applied, and the desired outcome/properties. From the large number of coarse-grained studies in the literature, especially in the field of computational biophysics and biology [9], very restricted representative examples are given here. CG is used for modeling polyelectrolytes (figure 7.2), protein folding (figure 7.3), knots in proteins (figure 7.4), and may other systems.

representation: all-atom   coarse-grained

chain of atoms   maps to   one bead
⇒

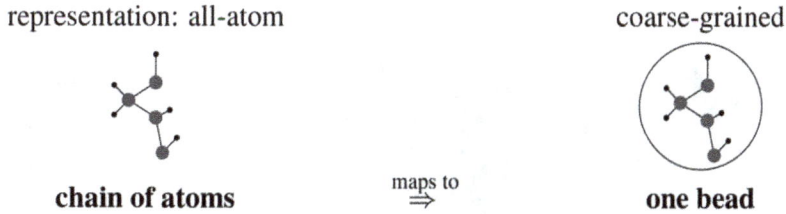

Figure 7.1. A typical CG of the all-atom picture in the left. All-atoms are represented by one bead in the coarse-grained representation on the right.

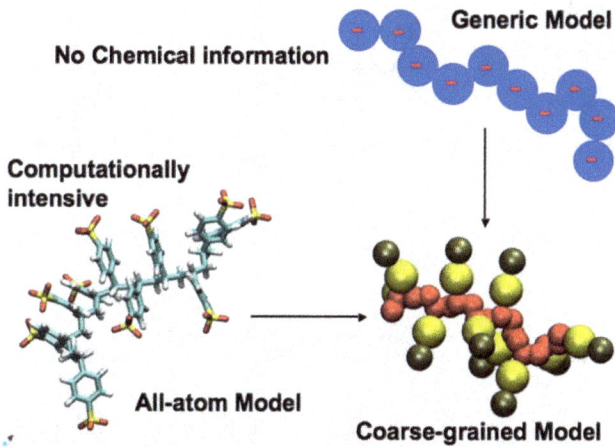

Generic Model

No Chemical information

Computationally intensive

All-atom Model   **Coarse-grained Model**

Figure 7.2. A CG procedure for polyelectrolytes: from the all-atom model, to a CG model, and a generic model where no chemical specificity is included. Courtesy of the Yethiraj group, Chemistry Department, University of Wisconsin-Madison.

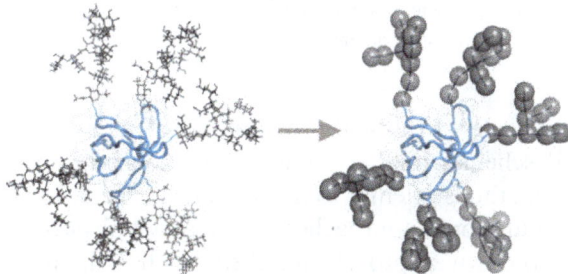

Figure 7.3. The effect of glycosylation on protein folding is examined through coarse-graining the protein [10]. Permission taken from the Proceedings of the National Academy of Sciences, USA.

7.2 Multiscale or hybrid schemes

Multiscale (MS) modeling is the field of performing computer simulations combining different temporal and spatial scales [11–14]. It can be applied to different fields in physics, biology, chemistry or engineering whenever the problems have important

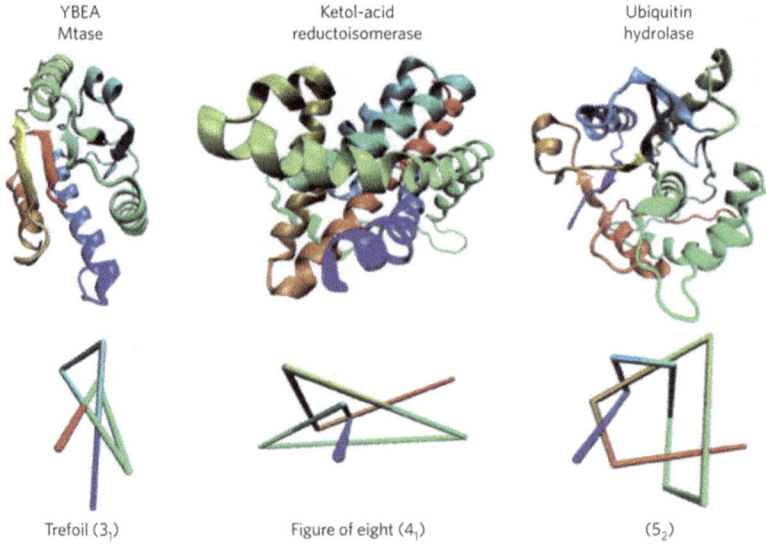

Figure 7.4. Knots in protein folding can be coarse-grained to rods without losing the essential information [15]. Permission taken from the Nature Publishing Group.

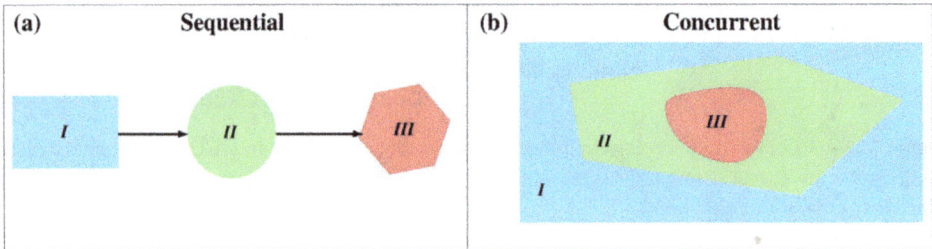

Figure 7.5. A sketch of (a) a sequential(serial) and (b) a concurrent(parallel) MS method. The three computational methods I, II, and III span different time and length scales. In (b) these should be seamlessly coupled at their boundaries.

features at different scales. Multiscale computational methods include more than one computational schemes and are thus often also named hybrid methods. In essence, what these methods attempt to do is to bridge the different scales shown in figure 1.2. MS computational approaches try to model physical systems through a bottom-up or a top-down approach sketched in that figure. In the first case, simulations begin at a very low and accurate level in order to reach the information of higher scales. In the second case, the process is reversed, simulations at higher scales lead to information on the microstructure and smaller scale.

In MS different approaches can be followed, as sketched in figure 7.5: (a) concurrent or parallel, in which different computational schemes are coupled within the same simulation and (b) sequential or vertical, in which results from one scheme complement those from a computational scheme at another level. New, faster algorithms are being developed in order to perform MS simulations especially for (a). In this way, novel medical and biological applications can be realized. In these methodologies it is always important to achieve an efficient mathematical treatment of the boundaries between the methods. In principle, a seamless coupling of the different computational methods at the boundaries (in time and length) where these 'touch' is needed. Typically, MS approaches are realized using analytical and numerical methods. For example, classical MS algorithms use the multi-grid method [16], the fast multiple method [17], the domain decomposition methods [18], the adaptive mesh refinement [19], the multi-resolution representation [20], etc.

7.2.1 Examples: sequential (serial) schemes

No specific methodologies for sequential MS modeling exist. In principle, any kind of computational method can be used in 'cooperation' with another method. The larger scale methods 'cooperate' with the lower scales by including information from the output of the lower scale in their inputs. Single-scale methods are all methods reviewed in the previous chapters, QM and electronic structure schemes, classical atomistic schemes like MD, semi-empirical, stochastic like MC or discretized schemes like finite elements or lattice Boltzmann, etc. The exchange of information among the schemes is being done 'externally', that is, not within the same algorithm. In this sense, no explicit coupling of the different approaches needs to be defined, as only the inputs and outputs of the methods are relevant to each other. Accordingly, the examples shown here are representative of the way sequential MS algorithms can perform. Such methodologies have been applied to fields like energy gas storage (figure 7.6), which use first-principles calculations to derive force fields in order to perform MC simulations on the adsorption performance of certain systems. An essential ingredient of concurrent MS schemes is the use of different temporal resolutions for each computational method. In a pioneering work in the late 1980s, high quality QM methods could evaluate the interaction among water molecules [21]. These interactions formed a database on which empirical potentials were parametrized and used in MD simulations, which were in turn used to evaluate the

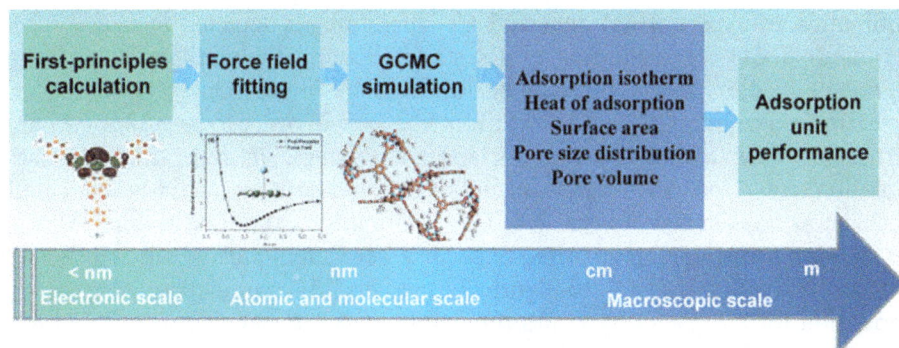

Figure 7.6. A sequential coupling of the scales involving first-principles and grand-canonical Monte Carlo (GCMC) simulations for the study of adsorptive processes for energy gas storage [22]. Reprinted with permission from the Royal Society of Chemistry.

A New Paradigm in Biomolecular Multiscale Simulation*

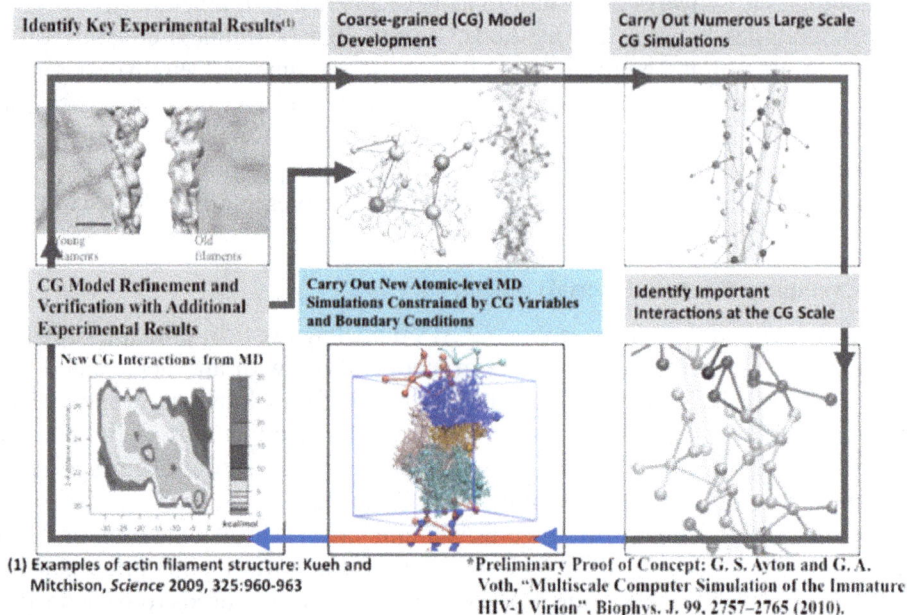

Figure 7.7. A paradigm for biomolecular MS simulations. CG simulations are carried out and refined based on experimental data to realize larger scale simulations, as denoted by the arrows. Courtesy of the Voth Lab, Chemistry Department, University of Chicago.

viscosity of water from the atomic autocorrelation function. This computed viscosity was used in a computational-fluid-dynamics calculation to predict tidal circulation in the sea.

A large number of biomolecular studies [9] are performed using MS methodologies as the chemical details of parts of the systems are typically of high importance, but the respective computational effort is too high and the systems too complex [23]. In biomolecular simulations, MS computer schemes, employing a combination of experimental data and CG methods, as depicted in figure 7.7 are often used. A CG representation of a system is assumed which is refined through experimental observations and can be followed by larger scale CG simulations. These can then be mapped back to the atomistic level in order to get a more accurate microscopic view of the system. As another example, a sequential multiscale simulation could provide important information about an actin-based cellular cytoskeleton, as sketched in figure 7.8. This information cannot be obtained through MD simulations, which are possible only for a very small part of the cytoskeleton, an actin subunit (panel (a) of figure 7.8). Such MD simulations can provide information for a higher length and time scale to parameterize a CG model of an actin filament. This will provide the input for a mesoscopic simulation of the cytoskeleton network (panel (d) of figure 7.8). Another example involves the electron localization along stretched DNA, which has been studied through electronic structure calculations of nucleobases at various distances and compared

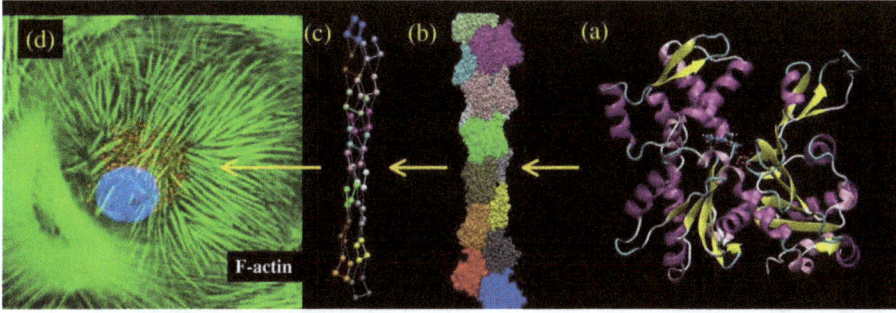

Figure 7.8. A visual representation of the MS challenge for understanding spatiotemporal coupling in biological systems: (a) an atomistic representation of an actin subunit in the monomeric state, (b) an actin filament, made up of many actin subunits, (c) a CG representation of an actin filament, and (d) a mesoscopic cytoskeleton network made up of many individual filaments. Courtesy of the Voth group, Chemistry Department, University of Chicago.

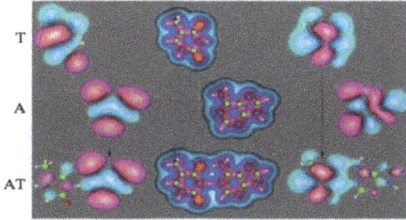

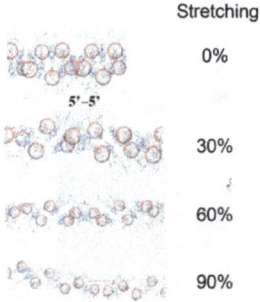

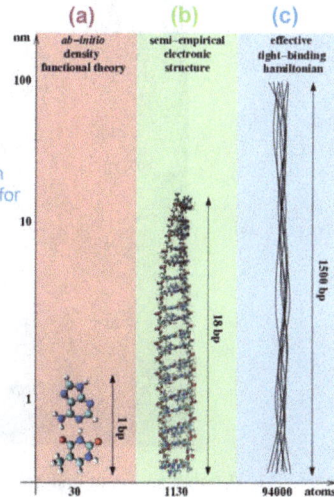

Figure 7.9. QM calculations (a) are compared to the outcome of a semi-empirical electronic structure method for short DNA strands in order to construct a Hamiltonian for investigating very long DNA strands (c) [24]. The panels of this figure are reproduced with permission from Springer.

to the electronic behavior of a semi-empirical electronic structure method of very short DNA strands. The outcome was in turn used to construct an effective Hamiltonian and model the electron localization along very long DNA molecules, as sketched in figure 7.9. Sequential MS techniques are also used to study thermo-electric materials and their phase diagram by combining an *ab initio*-based cluster

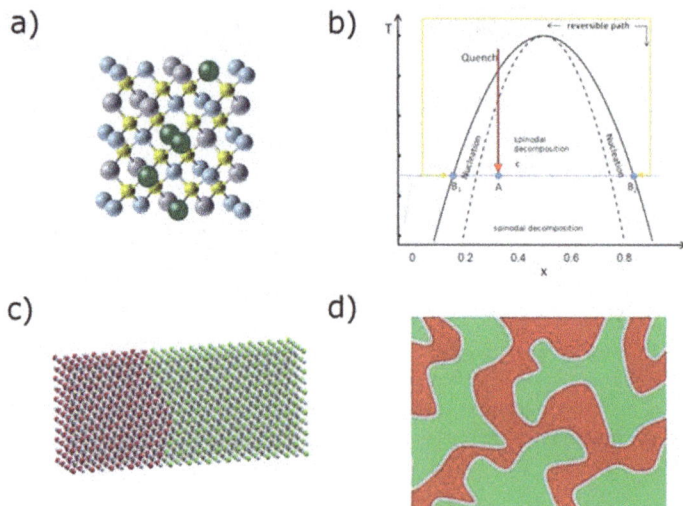

Figure 7.10. The phase diagram (b) of thermoelectric materials like Co(Ti, Mn)Sb can be studied by combining DFT simulations of a small part (a) of these materials with MC simulations (c) and phase-field theory (d) [25]. Courtesy of the Gruhn group, University of Bayreuth. Reproduced with permission from Wiley.

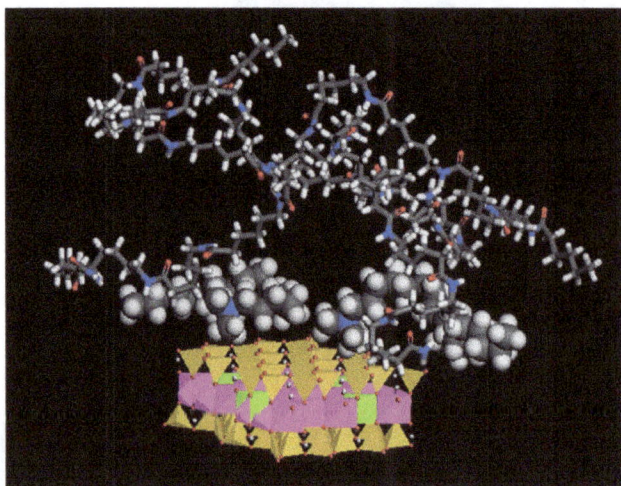

Figure 7.11. A combination of first-principles, dissipative particle dynamics (DPD), and finite elements (FEM) calculations has revealed the macroscopic properties of the polymer-clay nanocomposites shown [26]. Reproduced with permission from Elsevier.

expansion of the configurational energy with Monte Carlo and mean field calculations at higher temperature on a larger scale (figure 7.10). In a similar way, the behavior of polymer-clay nanocomposites has been studied combining quantum/force-field-based atomistic simulations to derive the interaction energies among all components. These are then mapped onto mesoscale dissipative particle dynamics (DPD) simulations followed by finite-element calculations (FEM). In the end, the macroscopic properties of the composites are obtained (figure 7.11).

7.2.2 Examples: concurrent (parallel) schemes

In contrast to the sequential MS techniques, certain methods have been developed for performing concurrent(parallel) MS simulations. These mainly focus on efficient ways to couple the separate regions (modeled through the different simulation schemes) in the simulations. Treating these boundaries between the regions of different computational methodologies is a very complex and delicate issue in MS simulations. The coupling of two (or more) methods is being done at two levels: (a) the information on the forces acting on particles is being exchanged across the scales and/or (b) the particles change their identity depending on which region they reside in. Regarding (a) the positions and forces are computed and are carefully distributed across the regions. Treating (b) is a bit more tricky and needs to be done in a sophisticated manner. The particles have a certain identity in each region, but need to change their identity when they move to another region modeled though another scheme.

For example, a coupled all-atom CG simulation may include the concept of a healing region, as depicted in figure 7.12. In such a scheme, the coarse-grained particles cannot suddenly switch to all-atom ones for which there is complete chemical specificity is taken into account. On the contrary, an intermediate healing region exists in which the particles change their identity [27]. The interaction between the particles is the superposition of the atomistic and coarse-grained potentials, where their weight is determined by the position of the interacting particles, as seen in figure 7.12. Another example is to link atomistic, coarse-grained,

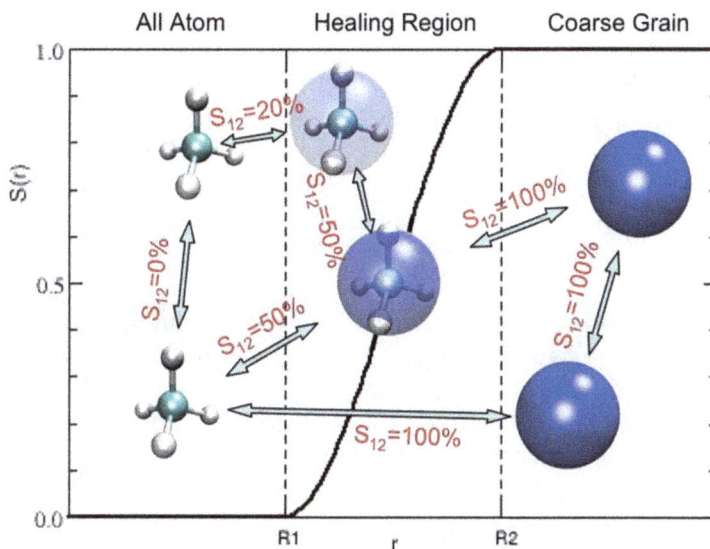

Figure 7.12. The all-atom and the coarse-grained regions are sketched. These are coupled though a healing region in which the particles change their identity ('all atom' or 'coarse-grain'). Courtesy of the Computational Chemistry Group, University of Amsterdam. Reproduced with permission from the Royal Society of Chemistry [27].

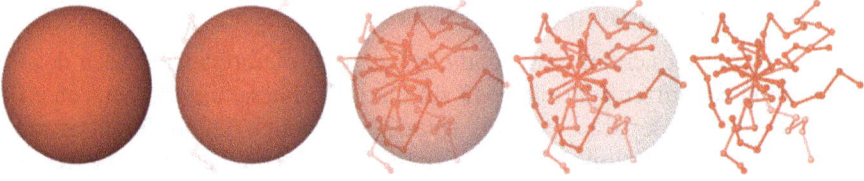

Figure 7.13. A schematic representation of an equilibration process of polymers. In the beginning (far left), each polymer is represented as a coarse-grained particle. During the equilibration process the atomistic degrees of freedom are gradually turned on, while the coarse-grained ones are gradually turned off. In the end, each polymer is modeled with the fine-grained representation (far right picture). Reproduced with permission from the European Physical Society [28].

and continuum (Eulerian) fluid dynamics in the general framework of fluctuating Navier–Stokes equations for molecular simulations of open domains with relatively large molecules [28]. As the dynamics proceed throughout the simulations, coarse-grained particles turn into atomistic ones and vice versa. Each time a coarse-grained particle approaches the atomistic region, the degrees of freedom of the atomistics need to be gradually turned on sharpening the resolution, as sketched in figure 7.13.

7.2.3 Quantum-mechanics molecular-mechanics (QM/MM)

A very common example of a concurrent multiscale (or hybrid) scheme is the QM/MM method, which is a combination of quantum mechanics (QM) and molecular mechanics (MM). QM/MM includes the accuracy of the QM schemes and the ability of MM to model large scales. MM is a force field method implicitly treating the electrons, does not include polarization or electron transfer and cannot model bond breaking/formation [29]. QM/MM combines both methods by partitioning the system into an electronically important region and the remainder, which acts as a perturbation and is treated classically [30]. In this way, the reacting part of the system is treated quantum mechanically, but the treatment of the less important region through MM significantly reduces the quantum mechanical part of the system, the computational time and the memory storage capacity. The combined QM/MM Hamiltonian for both parts of the systems would have the form:

$$H_{\text{tot}} = H_{\text{QM}} + H_{\text{MM}} + H_{\text{QM/MM}} \qquad (7.1)$$

where H_{QM} is the original Hamiltonian for the quantum-mechanical system as typically discussed in chapter 2, H_{MM} is the Hamiltonian based on the force field taken for MM, and $H_{\text{QM/MM}}$ includes the interaction between QM and MM. In principle, the total Hamiltonian of the system is a modification of the QM Hamiltonian presented in chapter 2 and reads:

$$H_{\text{tot}} = -\frac{1}{2}\sum_{I}\frac{1}{M_I}\nabla_I^2 - \frac{1}{2}\sum_{n}1/m_n\nabla_n^2 + \sum_{I<J}\frac{Z_I Z_J}{\left|\vec{R}_I - \vec{R}_J\right|} - \sum_{I,n}\frac{Z_I}{\left|\vec{R}_I - \vec{r}_n\right|}$$
$$+ \sum_{n<m}\frac{1}{\left|\vec{r}_m - \vec{r}_n\right|} - \sum_{Kn}\frac{Q_K}{\left|\vec{R}_K - \vec{r}_n\right|} + \sum_{K,L}\frac{Q_K Q_L}{\left|\vec{R}_K - \vec{R}_L\right|} \qquad (7.2)$$

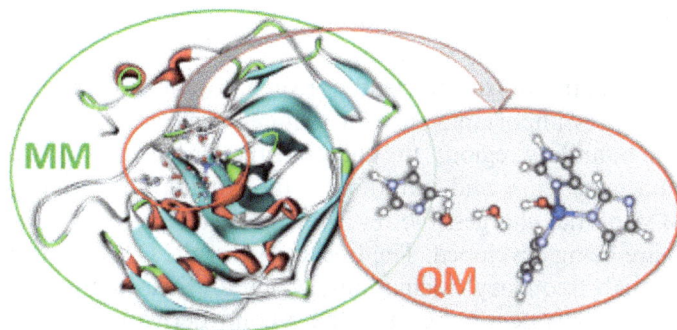

Figure 7.14. A typical separation of a biomolecular system treated with QM/MM (see text). Courtesy of the Sierka Lab, Faculty of Physics and Astronomy, Friedrich-Schiller-University Jena.

where K and L run over all MM atoms with charges Q. The two last terms in equation (7.2) map the interaction of electrons in QM with charges in MM and the electro- static interaction among MM atoms, respectively. A representative example of a QM/MM simulation is a biophysical system in figure 7.14. The small part, including a reaction process is captured with QM, while the much larger surrounding part of the molecule is treated with MM.

The above expressions can easily be implemented in simulations as long as the two sub-systems, the QM and the MM parts are well chosen and defined. What is also of very great importance for the efficiency of the QM/MM method is the treatment of the physical boundaries between the two QM and MM parts through the treatment of the bonds. A conventional solution to link the atoms of QM to MM at the boundaries is to add atoms along the bond crossing the boundaries (usually hydrogens, halogens or methyl groups). The link atom should satisfy the valence of the QM region. The QM atom is used for the calculation of all MM bonded terms. For the non-bonded (typically electrostatic) terms, the link atom does not interact with the MM atoms. Better properties can be achieved if the link atom interacts with the entire MM region, but in this way the electron density is poorly handled. On top of this, two improved bond treatments have been developed: (i) the local self-consistent field (LSCF) [31], which uses parameterized frozen orbitals along the QM/MM bond and are not optimized in the SCF cycle of QM and (ii) the generalized hybrid orbital (GHO) [32], which includes QM/MM orbitals in the SCF of the QM part.

Regarding the dynamics, chemical reactions are often simulated by MD, e.g. umbrella sampling together with the dynamics of a QM/MM system is almost identical to those of the MM part of the system. Within this framework, the forces are calculated from the first derivatives of the energy on each atom, the QM nuclei are treated identically to MM partial charges, and the system is being propagated by standard Newtonian dynamics. Note, also, that QM/MM can be used with MC. In those cases, the MM atoms affect the QM electron density and an SCF is required for all MC moves. An approximate energy change in the QM region is handled by a first-order perturbation theory. This results in a perturbative QM/MC, as long as the moves are far away from the QM region.

The QM/MM methods can efficiently treat a lot of systems in which bond formation/breaking is important, but still requires some parameterization for the treatment of the boundary regions, the choice of the QM size is not straightforward and depends on experience and intuition. Although the QM region polarizes in response to MM partial charges, the reverse is not true and fully polarizable QM/ MM methods are being developed. The free energy of the QM part of the system is determined through frequency calculations, which are inaccurate when applied to QM/MM systems. This occurs because the second derivatives are poorly determined, mainly due to the assumed harmonic approximation. Other approaches to the QM/MM exist. An example is the scheme ONIOM [33], which divides the system into the real (full) and the model (subset) parts, treated at a high and low level, respectively. Another method, known as the empirical valence bond [34] assumes any point on the reaction surface as a combination of two or more valence bond structures. The method takes a parameterization from QM or experiments and needs to be carefully set up, but is effective. Finally, the effective fragment potential [35], adds fragments to the standard QM treatment, it is fully polarizable and parameterized from separate *ab initio* calculations. However, the treatment of bonds between the true QM region and fragments of it still remain problematic.

7.2.4 Coupling the lattice Boltzmann method with MD

Another example of a concurrent MS scheme is the coupling of a mesoscopic fluid modeled through the lattice Boltzmann method with the dynamics of an object, like a polymer [36, 37]. In such a framework, the polymer modeled through a chain of beads is moving in a lattice resembling the presence of a fluid. The coupling of lattice Boltzmann and MD is being done through a frictional coefficient η, which is a tunable parameter. The information exchange between the two schemes is being done through the forces:

$$\vec{F}_p = -\eta\left(\vec{u}_p - \vec{U}_p\right) \qquad (7.3)$$

$\vec{F}_p$ is the force exerted by the fluid on the p-th monomer of the polymer and $\vec{U}_p$ is the local flow speed at the monomer position evaluated by a simple interpolation from the nearest grid points surrounding the monomer.

The method can be applied to various problems involving the motion of molecules in aqueous solutions. It has been applied in modeling hydrogels [38], the translocation of polymers resembling DNA molecules through nanopores [39], the aggregation and vesiculation of membrane proteins [40] or more complex processes like the blood flow through human arterioles [41]. For the latter, it is possible to characterize coronary artery plaque and the local coronary artery hemodynamic environment in a precise, three-dimensional manner and identify specific areas of high endothelial stress, plaque formation and its progression (figure 7.15). Such studies could lead to valuable prognostic information in the future.

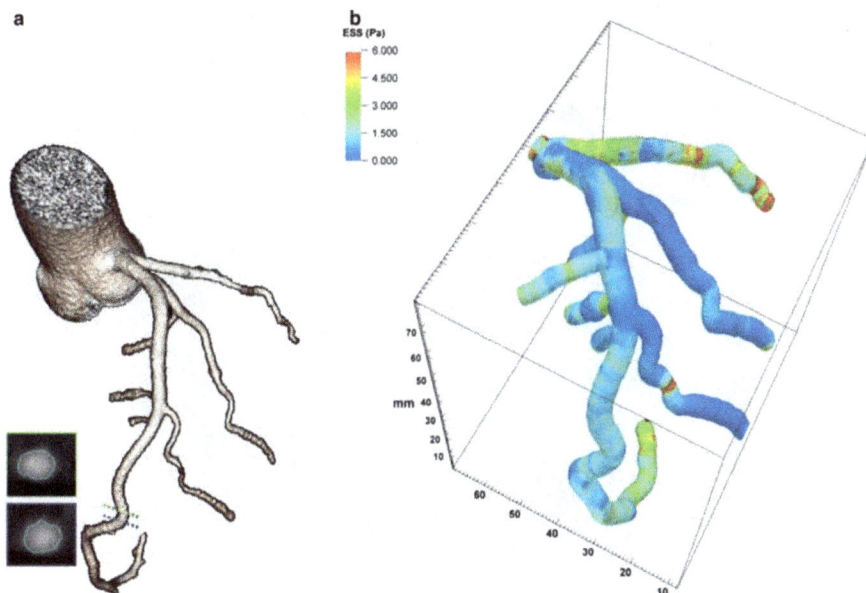

Figure 7.15. In (a) the automatically segmented coronary arteries and branches with lumen greater than or equal to 1 mm were obtained through computed tomography from a patient. In (b) the corresponding endothelial shear stress vascular profiling of the same coronary tree as in (a) using the coupled MD and Lattice Boltzmann scheme is depicted. The color coding shows the endothelial stress (EES) values reflected on the lumen surface, with blue depicting low EES and red depicting high EES. Reproduced from [41] with permission from Springer.

7.2.5 Other MS examples

In theory, concurrent MS approaches can couple any of the computational schemes, as long as the coupling is being done in a seamless way. This, though, is not always efficient. In many cases, a sequential coupling of schemes can lead to a more in-depth view of a physical system or process. For example, a simulation of crack propagation in silicon by seamlessly uniting quantum, atomistic, and continuum descriptions provides a quite accurate insight into the propagation and nucleation of dislocations in the material [42]. For this simulation, the geometrical decomposition of the silicon slab into five different dynamic regions is represented in figure 7.16. These are assigned to different methods and their boundaries ('hand-shaking' regions in which the exchange of information occurs): the continuum finite-element (FE) region, the atomistic molecular-dynamics (MD) region, the quantum tight-binding (TB) region, the FE–MD 'handshaking' region, and the MD–TB 'hand-shaking' region. As another example, it is also common to couple atomistics, like MD, with a mean field approach. For example, the atomic-level dynamics of Joule heating, melting, and plastic dynamics at Cu–Al metal contacts are modeled using an ad hoc coupling between a numerical solution to a heat transport equation, a virtual resistor network for describing electric current flow and a MD simulation using the embedded atom method (EAM) [43]. The different information provided from the two scales is depicted in figure 7.17. As a final example, parallel MS

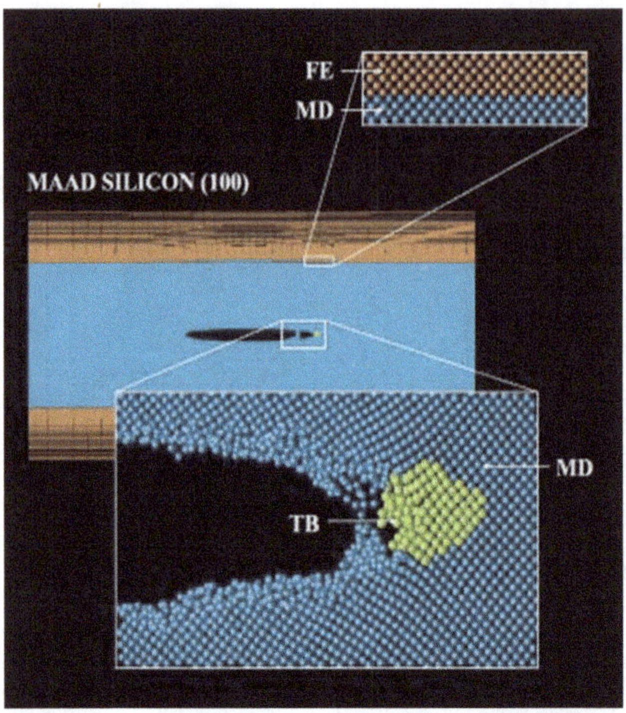

Figure 7.16. The image is the simulated silicon slab, with expanded views of the FE–MD (orange nodes and blue atoms) interface and the TB (yellow atoms) region surrounded by MD (blue) atoms. The TB region surrounds the crack tip with broken-bond MD atoms trailing behind this region. Only a proportion of the FE and MD regions is shown, since their extent is large. Reproduced from [42] with permission from the American Institute of Physics.

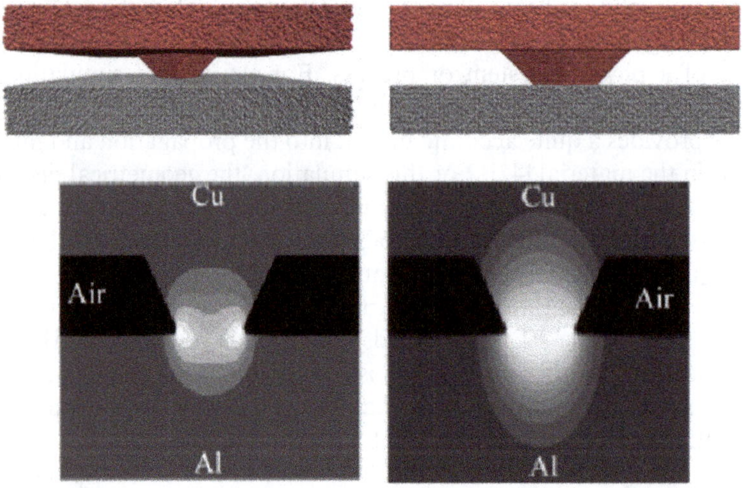

Figure 7.17. Top: 25 ps long MD simulations of loaded nanometer-scale Cu and Al asperity contacts. Bottom: temperature contours of the same material. Reproduced from [43] with permission from the Institute of Physics Publishing.

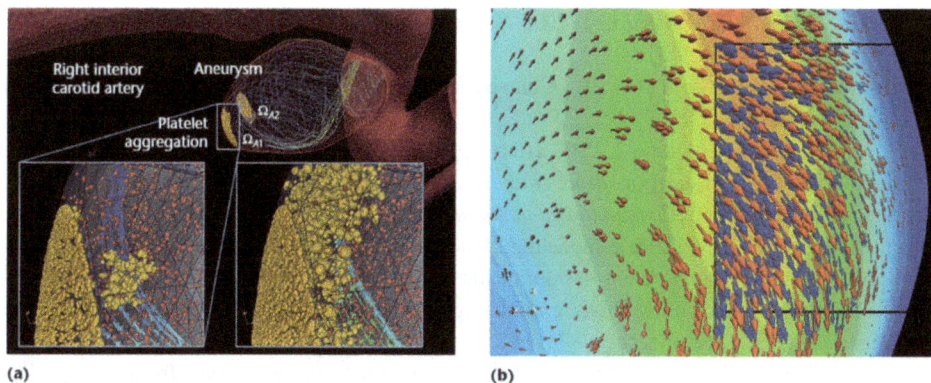

Figure 7.18. Left: streamlines and vectors depicting the instantaneous flow direction obtained from the Navier–Stokes solver. Right: brain vasculature, coupled continuum-atomistic simulation showing results on platelets aggregation on the wall of aneurysm with and without the glycocalyx separating the particles. The active platelets, inactive platelets, and glycocalyx, are shown in yellow, red and blue, respectively. The streamlines are also shown. The insets in (a) show clot formation (left) and platelets deposition (right). Reproduced from [44, 45] with permission from the IEEE and Elsevier.

simulations of a brain aneurysm are shown in figure 7.18. These are based on a computational scheme which involves coupling a high-order (spectral element) Navier–Stokes solver with a stochastic (coarse-grained) MD solver based on DPD, and has enabled the first MS simulation on 190 740 processors [44]. The Navier–Stokes approach gives an insight on the large scale flow features, while the results of the coarse-grained stochastic MD are related to the blood rheology inside the aneurysm, as inferred from figure 7.18. As a conclusion, it was not possible to include all the very important, innovative, and interesting MS and coarse-grained studies in this book. The interested reader is asked to look more deeply into the literature using as a starting point the studies presented here to find many more examples.

References

[1] Izvekov S and Voth G A 2005 A multiscale coarse-graining method for biomolecular systems *J. Phys. Chem.* B **109** 2469–73
[2] Voth G A 2008 *Coarse-Graining of Condensed Phase and Biomolecular Systems* (Boca Raton, FL: CRC Press)
[3] de Pablo J J 2011 Coarse-grained simulations of macromolecules: from DNA to nano-composites *Annu. Rev. Phys. Chem.* **62** 555–74
[4] Noid W G *et al* 2008 The multiscale coarse-graining method. i. a rigorous bridge between atomistic and coarse-grained models *J. Chem. Phys.* **128** 244114
[5] Feret J *et al* 2009 Internal coarse-graining of molecular systems *Proc. Natl Acad. Sci.* **106** 6453–8
[6] Gohlke H and Thorpe M F 2006 A natural coarse graining for simulating large biomolecular motion *Biophys. J.* **91** 2115–20
[7] Padding J T and Louis A A 2006 Hydrodynamic interactions and brownian forces in colloidal suspensions: coarse-graining over time and length scales *Phys. Rev.* E **74** 031402

[8] Pivkin I V and Karniadakis G E 2008 Accurate coarse-grained modeling of red blood cells *Phys. Rev. Lett.* **101** 118105

[9] Saunders M G and Voth G A 2013 Coarse-graining methods for computational biology *Ann. Rev. Biophys.* **42** 73–93

[10] Shental-Bechor D and Levy Y 2008 Effect of glycosylation on protein folding: a close look at thermodynamic stabilization *Proc. Natl Acad. Sci.* **105** 8256–61

[11] Steinhauser M O 2007 *Computational multiscale modeling of fluids and solids* (Berlin: Springer)

[12] Baeurle S A 2009 Multiscale modeling of polymer materials using field theoretic methodologies: a survey about recent developments *J. Math. Chem.* **46** 363–426

[13] Horstemeyer M F 2009 Multiscale modeling: a review *Practical Aspects of Computational Chemistry* (Berlin: Springer) pp 87–135

[14] Peter C and Kremer K 2009 Multiscale simulation of soft matter systems-from the atomistic to the coarse-grained level and back *Soft Matter* **5** 4357–66

[15] Shakhnovich E 2011 Protein folding: to knot or not to knot? *Nat. Mater.* **10** 84–6

[16] Hackbusch W 2013 Multi-grid methods and applications Vol. 4 (Berlin: Springer)

[17] Katoh K *et al* 2002 MAFFT: a novel method for rapid multiple sequence alignment based on fast Fourier transform *Nucleic Acids Research* **30** 3059–66

[18] Brezzi F and Luisa D M 1994 A three-field domain decomposition method *Contemporary Mathematics* **157** 27–27

[19] Berger M J and Oliger J 1984 Adaptive mesh refinement for hyperbolic partial differential equations *J. Comput. Phys.* **53** 484–512

[20] Natrajan A, Reynolds P F and Srinivasan S 1997 MRE: a flexible approach to multi-resolution modeling Parallel and Distributed Simulation, 1997 *Proc. 11th Workshop on IEEE*

[21] Clementi E and Reddaway S F 1988 Global scientific and engineering simulations on scalar, vector and parallel LCAP-type supercomputers [and discussion] *Phil. Trans. R. Soc. A* **326** 445–70

[22] Xiang Z *et al* 2010 Multiscale simulation and modelling of adsorptive processes for energy gas storage and carbon dioxide capture in porous coordination frameworks *Energy Environ. Sci.* **3** 1469–87

[23] Marée A F M *et al* 2006 Polarization and movement of keratocytes: a multiscale modelling approach *Bull. Math. Biol.* **68** 1169–211

[24] Barnett R L, Maragakis P, Turner A, Fyta M and Kaxiras E 2007 Multiscale model of electronic behavior and localization in stretched dry DNA *J. Mater. Sci.* **42** 8894–903

[25] Mena J *et al* 2016 Nanophase separation in CoSB-based half-Heusler thermoelectrics: A multiscale simulation study, *Phys. Status Solidi A* **213** 706–15

[26] Scocchi G *et al* 2007 To the nanoscale, and beyond!: multiscale molecular modeling of polymer-clay nanocomposites *Fluid Phase Equilib.* **261** 366–74

[27] Nielsen S O, Bulo R E, Moore P B and Ensing B 2010 Recent progress in adaptive multiscale molecular dynamics simulations of soft matter *Phys. Chem. Chem. Phys.* **12** 12401–14

[28] Delgado-Buscalioni R, Sablić J and Praprotnik M 2015 Open boundary molecular dynamics *Eur. Phys. J. Spec. Top.* **224** 2331–49

[29] Burkert U *et al* 1982 *Molecular Mechanics* vol 177 (Washington, DC: American Chemical Society)

[30] Warshel A and Levitt M 1976 Theoretical studies of enzymic reactions: dielectric, electrostatic and steric stabilization of the carbonium ion in the reaction of lysozyme *J. Mol. Biol.* **103** 227–49

[31] Monari A, Rivail J-L and Assfeld X 2012 Theoretical modeling of large molecular systems. advances in the local self consistent field method for mixed quantum mechanics/molecular mechanics calculations *Acc. Chem. Res.* **46** 596–603

[32] Gao J, Amara P, Alhambra C and Field M J 1998 A generalized hybrid orbital (GHO) method for the treatment of boundary atoms in combined qm/mm calculations *J. Phys. Chem.* A **102** 4714–21

[33] Svensson M *et al* 1996 Oniom: A multilayered integrated MO + MM method for geometry optimizations and single point energy predictions. a test for Diels–Alder reactions and Pt (P (t-Bu) $_3$)$_2$ + H$_2$ oxidative addition *J. Phys. Chem.* **100** 19357–63

[34] Kamerlin S C L and Warshel A 2011 The empirical valence bond model: theory and applications *WIREs: Comput. Mol. Sci.* **1** 30–45

[35] Gordon M *et al* 2001 The effective fragment potential method: a QM-based MM approach to modeling environmental effects in chemistry *J. Phys. Chem.* A **105** 293–307

[36] Ahlrichs P and Dünweg B 1998 Lattice-Boltzmann simulation of polymer-solvent systems *Int. J. Mod. Phys.* C **9** 1429–38

[37] Ahlrichs P and Dünweg B 1999 Simulation of a single polymer chain in solution by combining lattice boltzmann and molecular dynamics *J. Chem. Phys.* **111** 8225–39

[38] Mann B A F, Kremer K, Lenz O and Holm C 2011 Hydrogels in poor solvents: a molecular dynamics study *Macromol. Theory Simul.* **20** 721–34

[39] Fyta M G, Melchionna S, Kaxiras E and Succi S 2006 Multiscale coupling of molecular dynamics and hydrodynamics: application to DNA translocation through a nanopore *Multiscale Model. Simul.* **5** 1156–73

[40] Reynwar B J, Illya G, Harmandaris V A, Müller M M, Kurt Kremer and Deserno M 2007 Aggregation and vesiculation of membrane proteins by curvature-mediated interactions *Nature* **447** 461–4

[41] Rybicki F J *et al* 2009 Prediction of coronary artery plaque progression and potential rupture from 320-detector row prospectively ECG-gated single heart beat CT angiography: Lattice Boltzmann evaluation of endothelial shear stress *Int. J. Cardiovascular Imaging* **25** 289–99

[42] Abraham F F, Broughton J Q, Bernstein N and Kaxiras E 1998 Spanning the length scales in dynamic simulation *Comput. Phys.* **12** 538–46

[43] Irving D L, Padgett C W and Brenner D W 2008 Coupled molecular dynamics/continuum simulations of joule heating and melting of isolated copper–aluminum asperity contacts *Model. Simul. Mater. Sci. Eng.* **17** 015004

[44] Grinberg L *et al* 2011 A new computational paradigm in multiscale simulations: application to brain blood flow in *SC '11 Proceedings of 2011 International Conference for High Performance Computing, Networking, Storage and Analysis, ACM, New York* doi:10.1145/2063384.2063390

[45] Grinberg L, Fedosov D A and Karniadakis G E 2013 Parallel multiscale simulations of a brain aneurysm *J. Comput. Phys.* **244** 131–47

Chapter 8

Other common aspects

The previous chapters reviewed the core methods used in physics. There are additional computational methods, which are mainly used in other disciplines and are not reviewed here. Other schemes in physics also exist and are being implemented in a variety of problems. In this chapter, focus will be given to methods for calculating free energy differences, schemes to add electrostatic and electrokinetic effects, and some other tools for computing in physics. As a first step, however, an important aspect in computer simulations in physics, that is, the way to search for the global minimum of a system, will be tackled.

8.1 Search and sampling of the configuration space, energy minimization

In MD (and also MC) simulations searching for the global energy minimum of a system are essential for the determination of the properties. This 'search problem' cannot always be efficiently solved. A system can have a very large number of degrees of freedom related to harmonic, non-harmonic, chaotic, or diffusive motion. Correlations are almost always present and cover time and spatial scales from fs and nm to ms and μm. In this respect, the potential function defines a very rugged energy hyper-surface. The energy surface has energy wells and mountains of a wide range of depths and heights and spatial extent. Accordingly, searching the regions of the system, which mostly contribute to the free energy (looking for a global minimum of a high-dimensional function) is difficult. A global minimum can be described only by a statistical mechanics ensemble of configurations in which the weight is given by the Boltzmann factor $P(x) \propto \exp\left(-\frac{V(x)}{k_B T}\right)$, where T is the temperature, $V(x)$ the potential energy, and x is a characteristic parameter. The equilibrium properties of the system are dominated by parts of configuration space having a low $V(x)$. High-energy regions of the energy hyper-surface do not contribute configurations relevant to the state of the system, unless they are too many.

doi:10.1088/978-1-6817-4417-9ch8
8-1

Search techniques depend on the potential $V(x)$, as well as the number and types of degrees of freedom. The 'probability search' techniques also depend on the Boltzmann probabilities $P(x)$. The basic search methods are either *systematic* or *heuristic*. (a) **Systematic or (exhaustive)** search methods scan the complete (or a significant fraction of the) configuration space of the physical system. Excluded are particular subspaces which are expected not to contain the desired solution only if there is no reduction in the quality of the result. Exclusion is usually based on *a priori* (physical/chemical nature) knowledge about the structure of space or the energy hyper-surface. These algorithms are typically applied to small systems because of the exponential growth of the computing time as the degrees of freedom increase. (b) **Heuristic** search methods aim to generate a representative (according to Boltzmann weighting) set of system configurations by visiting a very small fraction of the configuration space. There are three types of heuristic schemes: (i) **non-step** which generate a series of independent system configurations, (e.g. the distance-geometry metric matrix method, involving uncorrelated series of random configurations cast into a distance-based form). (ii) **Step**, which build a complete system of configurations of system fragments in a step-wise manner (e.g. build-up Gibson–Scheragen, a combinatorial build-up using dynamic programming techniques. (iii) **Step methods**, which generate a new configuration of the complete system from previous configurations (e.g. energy minimization, Metropolis MC, MD, stochastic dynamics). These methods are classified with respect to the way the step direction and step size are chosen.

Representative examples of step methods are presented next. (a) **Energy minimization:** this is based only on energy values and random steps (**simplex** schemes) or on energy and energy gradient values (**steepest descent** and **conjugate gradient** schemes) or on the second derivatives of energy (**Hessian matrix** schemes). (b) **MC:** this randomly chooses a step direction, during which the step size is limited by the Boltzmann acceptance criterion (for a change in the system energy $\Delta V < 0$ the step in configuration space is accepted [1], for $\Delta V < 0$, the step is accepted with a probability $e^{-\Delta V/k_B T}$). (c) **MD:** the step is determined by the force $\left(\propto -\frac{\partial V}{\partial x}\right)$ and the inertia of degrees of freedom. The method is related to a short-time memory of the path followed until the actual step. (d) **Stochastic dynamics (SD):** here, a random component is added to the force. The size of this component is determined by the system temperature, atomic masses, and friction coefficients. In general, search schemes based on stepping through the configuration space using a combination of the above (energy, gradient, Hessian, memory, randomness) have also been developed.

The efficiency of the search methods is restricted by the functional form of the potential $V(x)$ to be explored in order to find the low-energy regions. The occurrence of high-energy barriers between local minima means that the convergence of the step methods is small. As a result it is necessary to enhance the search and sampling power of search methods, through **sampling enhancement techniques** such as:

- **Deformation or smoothing of potential energy surface:** here, the repulsive short-range interaction between overlapping atoms is softened to reduce energy barriers. The applied constraints freeze the highest-frequency degrees of freedom. The deformation of the energy surface during a simulation based on the diffusion equation is made proportional to the local curvature (second derivative) of the surface. Once a local minimum is found, it is removed from the potential energy surface through a suitable local deformation of the energy surface.

- **Scaling system parameters:** this scheme involves temperature annealing and atomic mass scaling and is a mean-field approximation.
- **Multi-copy search and sample:** these are genetic algorithms in which configurations are created and deleted. Examples are **replica-exchange** (multiple system copies at different temperatures) and **SWARM** (combines a collection of copies of a system each with its own trajectory into a cooperative multi-copy system which searches the configurational space).

The most common step methods relevant to the **energy minimization** of a system in a simulation are briefly reviewed in the following. These algorithms for geometry optimization of periodic or finite systems usually follow the path for minimizing the energy until convergence (within a given accuracy) is reached. Typical schemes are the steepest descent scheme, conjugate gradient, Newton–Raphson, Hessian, BFGS, L-BFGS, etc. The **steepest descent (SD)** scheme attempts to locate the minimum of a functional $F(x)$, which is assumed to have a minimum [2]. In order to find the local minimum of the function, steps proportional to the negative of the gradient at a given point are taken. Taking steps proportional to the positive gradient will lead to the maximum of the function. SD performs rather poorly when the minimum of $F(x)$ lies in a long narrow valley, as the direction of each step is orthogonal to the previous one. This would lead to a large number of iterations, but can be overcome by the **conjugate-gradient** scheme. Congugate-gradient is used to minimize the sum of eigenvalues of the problem [3]. One minimization step is made independent of another in order not to move errors along during the minimization process. The initial minimization direction is the negative of the gradient at the starting point. The subsequent conjugate direction is constructed from a linear combination of the new gradient and the previous direction, which minimized $F(x)$. Instabilities due to large super cell sizes, large plane-wave kinetic energy cutoffs can occur. A workaround could be achieved through **preconditioning**, that is, through multiplication of the steepest-descent vector by a preconditioning matrix to produce a preconditioned steepest-descent vector parallel to the error vector. It represents more accurately the error vector as the preconditioned vector is parallel to the error vector. It shows poor convergence as each step tends to remove components of the error vector corresponding to states in a particular energy range. The convergence is improved if the method conjugates for weighting factors which distinguish between the error vector and the steepest-descent vector. This is mainly used in QM schemes.

Overall, SD is a very robust method, works for any number of dimensions, performs a line search to find locally the optimal step size, but shows slow convergence towards the end. Many iterations are needed to get the local minimum with the required accuracy if the curvature in different directions differs significantly. In that case, it is necessary to use preconditioning which changes the geometry of the space to shape the function level sets. There is also no guarantee that the minimum will be reached in a finite number of iterations. Other methods may converge in fewer iterations, but have a higher computational cost at each iteration. In conjugate-gradient, the initial direction is negative to the gradient at starting point. The subsequent conjugate direction is constructed from a linear combination of the new gradient and previous direction which minimized $F(x)$. The search direction is generated using information about the function, as obtained from all sampling points along the conjugate-gradient path. Instabilities can occur for large supercell sizes, etc and preconditioning can be used to overcome these. Additional information on energy minimization can be found in the literature [4–6].

8.2 Free energy methods

A common target in simulations is the evaluation of thermodynamics properties. From these, the free energy (typically Helmholtz or Gibbs) is the most relevant one. Investigation of the free energy can lead to critical phenomena, phase transitions or other transformations. Typically, there is a difference between the computational techniques used to study for example first-order transitions from those used to study continuous phenomena and higher-order transitions [7–9]. As it is also not possible to calculate absolute values for the free energies[1], relative energies or differences in free energies are searched for. A quantitative comparison of such methods can be found elsewhere [10]. Free energy methods cover two topics: (i) the function of a state (A, G, Ω, ...) and its change during a process (ΔA, ΔG, $\Delta \Omega$, ...), and (ii) the profiles of the free energy with respect to an internal constraint, such as the reaction coordinate.

8.2.1 Free energy differences

According to statistical mechanics, assuming a Hamiltonian H, the Helmholtz free energy can be obtained by evaluating integrals of the type:

$$F = k_{\mathrm{B}}T \ln \left(\iint \mathrm{d}\vec{p}^{N} \, \mathrm{d}\vec{r}^{N} \exp\left[-\beta H\left(\vec{p}^{N}, \vec{r}^{N}\right)\right] \right) \tag{8.1}$$

In the simulations, such as MD or MC, high energy regions are not adequately sampled, making the evaluation of such integrals very difficult. Although absolute free energies are not easily computed, this is not the case for free energy differences between two states X and Y. The Helmholtz free energy difference between these two states will be

$$\Delta F = k_{\mathrm{B}}T \ln \left(\frac{\iint \mathrm{d}\vec{p}^{N} \, \mathrm{d}\vec{r}^{N} \exp\left[-\beta H_{Y}\left(\vec{p}^{N}, \vec{r}^{N}\right)\right]}{\iint \mathrm{d}\vec{p}^{N} \, \mathrm{d}\vec{r}^{N} \exp\left[-\beta H_{X}\left(\vec{p}^{N}, \vec{r}^{N}\right)\right]} \right) \tag{8.2}$$

---

[1] The free energy of a system is not a simple function of the phase space coordinates of the system, but a function of the Boltzmann-weighted integral over the phase space (a partition function) and the absolute free energies of two states cannot be calculated directly.

which holds also for reversing X and Y. Simulations should be performed to collect statistics for the states X and Y. In cases where these states do not overlap sufficiently, other intermediate states can be assumed to improve the degree of overlap and improve the sampling. The way this can be implemented has its own complications [7, 11, 12].

8.2.2 Free energy perturbation

A method to calculate free energy differences from MD or MC simulations is the free energy perturbation (FEP) [13]. In order to obtain the difference in the free energy from moving from one state (X) to another state (Y), the following expectation value needs to be computed over a simulation for state X:

$$\Delta F(X \rightarrow Y) = F_B - F_X = -k_B T \ln \left\langle \exp\left(-\frac{E_Y - E_X}{k_B T} \right) \right\rangle_X \tag{8.3}$$

In this equation, $k_B T$ is the thermal energy, while E_X and E_Y are the energies corresponding to states X and Y. Note, that the difference between the two states X and Y should be small enough, so that the method converges. It is also possible to divide the simulations into separate windows, when the difference is larger. In any case, while running a simulation for the starting state X, the energy for Y is also at some point sampled.

8.2.3 Thermodynamic integration

A very common and efficient scheme for calculating free energy differences is known as thermodynamic integration (TI). In TI, the free energy difference is calculated through the definition of a thermodynamic path between the relevant states X and Y and an integration over ensemble-averaged enthalpy changes along this path. In order to compute the free energy of a system at given conditions typically a reversible path (which is not necessarily physical) is taken, which links the state under consideration (Y) to a state of known free energy (X). The change in the free energy can then be evaluated by thermodynamic integration, that is, by integrating the equations which connect the free energy derivatives with a certain property of the system.

For a system with N particles and a potential energy U, take two systems X and Y, which have potential energies U_X and U_Y. These can be obtained through calculation of ensemble averages in MD or MC. Typically TI assumes a potential energy function which depends linearly on a coupling parameter λ, such that:

$$U(\lambda) = U_X + \lambda(U_Y - U_X) \tag{8.4}$$

For $\lambda = 0$, the potential function corresponds to the reference state X, while $\lambda = 1$ recovers the potential function of the state of interest Y. The partition function of the system in the canonical ensemble is defined as:

$$Q(N, V, T, \lambda) = \frac{1}{\Lambda^{3N} N!} \int d\vec{r}^N \exp\left[-\beta U(\lambda) \right] \tag{8.5}$$

The partition function can then lead to information on the free energy F of the system and the derivative of the free energy

$$\left(\frac{\partial F(\lambda)}{\partial \lambda}\right)_{N,V,T} = -k_{B}T\frac{\partial}{\partial \lambda}\ln Q(N, V, T, \lambda) = ...=\left\langle\frac{\partial U(\lambda)}{\partial \lambda}\right\rangle_{\lambda} \quad (8.6)$$

an ensemble average of the derivative of the potential energy with respect to λ. The change in the free energy between the systems X and Y can be obtained through an integration of equation (8.6) with respect to the coupling parameter λ:

$$\Delta F(\lambda = 0 \rightarrow 1) = F(\lambda = 1) - F(\lambda = 0) = \int_{\lambda=0}^{\lambda=1} d\lambda \left\langle\frac{\partial U(\lambda)}{\partial \lambda}\right\rangle_{\lambda} \quad (8.7)$$

and can be directly obtained in a simulation. In practice, a potential energy function $U(\lambda)$ needs to be defined, which samples the ensemble of equilibrium configurations at various λ values. Thermodynamic integration is possible for any non-linear potential function $U(\lambda)$, as long as $U(\lambda = 0) = U_X$ and $U(\lambda = 1) = U_Y$, and $U(\lambda)$ is differentiable. The linear interpolation of equation (8.4) is a good choice, as the sign of $\frac{\partial^2 F}{\partial \lambda^2}$ is known. In general, both intrinsically dynamic or static (like the one discussed) schemes for TI exist [14].

8.2.4 Umbrella sampling

Within the umbrella sampling framework discussed in chapter 6, the unbiased free energy profile $F_i(\xi)$ is obtained through the probability distribution of the system along the reaction coordinate, by eliminating all other degrees of freedom but ξ (equation (2.7)). It can be shown that $F_i(\xi)$ can be straightforwardly obtained through [15]

$$F_i(\xi) = -\frac{1}{\beta}\ln\left[P_i^b(\xi)\right] - w_i - \frac{1}{\beta}\ln\left\langle \exp\left[-\beta w_i(\xi)\right]\right\rangle \quad (8.8)$$

The derivation of $F_i(\xi)$ is exact and equation (8.8) is sufficient to unbias the simulations. As long as the choice of the bias potentials was good, the sampling is also assumed sufficient. In cases in which more windows need to be taken, free energies curves $F_i(\xi)$ of more windows need to be combined in finding a global $F(\xi)$. Different methods have been applied to combine these results. As an example, the **weighted histogram analysis method (WHAM)** is a very common scheme to analyze series of umbrella sampling simulations, that is, combine the windows taken along the reaction coordinate [16]. Overall, care needs to be taken when applying the umbrella sampling method and the exact choice of the windows, bias potentials, and all other details for efficiently implementing this scheme [7].

8.2.5 Metadynamics

Metadynamics is useful in cases where ergodicity is hindered by the form of the energy landscape of the system [17, 18]. The algorithm attempts to reconstruct the free energy (typically multidimensional) of systems and is usually implemented in MD schemes. The scheme is based on including artificial dynamics in sampling the phase space. The artificial dynamics is driven by the free energy and is biased by a history-dependent potential constructed as a sum of Gaussians centered along the trajectory of the collective variables S. These provide a coarse-grained description of the system.

During a simulation, the location of the system in space defined through S is determined and a positive Gaussian potential is added to the real energy landscape of the system. Additional Gaussians are continuously added discouraging the system to go back to its previous steps, until the system explores the full energy landscape. If a new Gaussian is added at every time interval t_G, the biasing potential at time t is defined as

$$V_G(S(x), t) = \sum_{t=t_G, 2t_G, 3t_G} w \exp\left(-\frac{(G(x) - s_{t_i})^2}{2\delta s^2}\right) \tag{8.9}$$

where w and δs are the height and the width of the Gaussians and $s_t = S(x(t))$ is the value of the collective variable at time t. This potential, which is a function of time, fills the minima in the free energy surface. As a result, the sum of the Gaussians and of the free energy becomes approximately a constant as a function of S, which is the reason for the collective variables to start fluctuating. At this point, the energy landscape can be recovered as the opposite of the sum of all Gaussians.

In metadynamics it is essential to choose the correct collective variables. Typically, these are related to stable minima in space and depend on the system and process to be studied [18]. Metadynamics can be exploited not only for efficiently computing the free energy, but also for exploring new reaction pathways [19] and accelerating rare events, investigating phase transitions [20], in protein folding [19], molecular docking [21], etc. A very common implementation of metadynamics is done in PLUMED [22], which offers interfaces for many of the commonly used MD programs (NAMD, Gromacs, LAMMPS, CP2k, etc).

8.2.6 Other free energy methods

Computationally faster methods than calculating thermal properties as discussed previously rely on the calculation of the differences in the chemical potential between given species and the ideal gas under the same conditions. This difference is also known as the excess chemical potential, for the calculation of which different methods have been developed, such as the **particle (Widom) insertion** method [23] or the **overlapping distribution** method [24]. In principle, many additional alternatives to the ones discussed have been developed. Examples of these are the **accelerated dynamics** methods [25], the local elevation scheme or **adaptive bias potential** [26], the **flooding potential** [27], the **nudge elastic band** methods [28], the **blue moon ensemble** theory [29], **multiple histograms** [30], **string methods** [31], the **pattern search** [32], **acceptance ratio** method [24], etc. All these techniques assume that the system under study is in a thermodynamic equilibrium or that any changes in the equilibrium state of the system happen very slow in time. In addition, non-equilibrium free energy methods have been developed and typically relate the free energy difference between two systems with the non-equilibrium work needed to transform one system to the other in a short time. Most common non-equilibrium free energy methods are the **Jarzynski equality** [33, 34] and its generalizations [35, 36].

8.3 Dealing with electrostatics/electrokinetics

A very common simplification in simulating physical systems is to assume that these are neutral. This is pretty efficient in terms of the computational effort that is being gained and the accuracy acquired. This is more often the case in quantum mechanical calculations. Nevertheless, in many cases, charges are essential, for example when ions are present and play an important role in defining the properties of the system.

8.3.1 Charges and electrostatic interactions

A first issue is the way to handle the presence of charges and their long-range interactions (Coulombic and dipolar). Calculating these is computationally more demanding than simple (short-range) bonded interactions. These vanish after some distance further away from the simulated particles, but the long-range ones need to be computed at larger distances as well. The short-range potentials are often being truncated in the simulations. Typically, this cannot be done for the long-range potentials as it can be shown that truncation plus tail correction for Coulombic and dipolar interactions leads to divergence in the tail correction of the potential [37]. In order to overcome this, a variety of schemes for efficiently handling long-range interactions have been developed. Most common schemes are the Ewald summation [38] and its variations, the particle-particle/particle-mesh (P^3m), ICC* etc. Further details and schemes can be found in the literature [39–47].

8.3.2 Ewald summation

One of the most common methods used for the calculation of long-range interactions in periodic, as well as finite [48] systems is the Ewald summation [38]. The method replaces the summation of interaction energies in real space with an equivalent summation in Fourier space and the long-range interactions are divided into two parts: a short-range contribution calculated in real space, and a long-range contribution which does not have a singularity and is calculated using a Fourier transform. As a result, the Ewald summation method converges faster than a direct summation and is very accurate. In the following, only the main ingredients of the Ewald summation evaluation are presented. More details can be found in the literature [7].

The system of interest includes positively and negatively charged particles with charges q_i, but is electrically neutral ($\sum_i q_i = 0$). All particles N are located in a cubic box of length L. The Ewald summation then computes the Coulomb contribution U_{Coul} to the potential energy of this system:

$$U_{\text{Coul}} = \frac{1}{2}\sum_{i=1}^{N} q_i \phi(r_i) \tag{8.10}$$

where $\phi(r_i)$ the electrostatic potential at the position of particle i with charge q_i. In principle, what needs to be done is to calculate the energy of a given charge distribution in Fourier space, which corresponds to the solution of Poisson's equation for the electrostatic potential. A key ingredient here is to take into acount the overcounting in the calculation of the charge distribution and the additional interaction between a continuous Gaussian charge cloud of charge q_i and a point charge q_i located at the center of the Gaussian should be corrected for. This is known as the spurious contribution (or self-interaction term) to the potential energy. The term is constant in a simulation if all charges of the system are fixed and needs to be subtracted from the long-range part of the interaction.

Putting together all the parts above leads to the total long-range contribution (equation (8.10)) within the Ewald sum approach:

$$\begin{aligned}
U_{\text{Coul}} &= \frac{1}{2V}\sum_{k\neq 0}\frac{4\pi}{k^2}\left|\rho\left(\vec{k}\right)\right|^2 \exp(-k^2/4\alpha) - \left(\frac{\alpha}{\pi}\right)^{1/2}\sum_{i=1}^{N} q_i^2 \\
&+ \frac{1}{2}\sum_{i\neq j}^{N}\frac{q_i q_j \text{erfc}\left(\sqrt{\alpha}r_{ij}\right)}{r_{ij}}
\end{aligned} \tag{8.11}$$

where erfc is the complementary error function, α is related to the width of the Gaussians, and ρ is the cloud charge density. Based on the expression of equation (8.11), it is also possible to find an expression for more complex cases, like a system containing dipolar molecules, or including the dielectric constant of a polar fluid. Care should also be given in treating the boundary conditions in the system [7]. The scaling of the method is $\mathcal{O}(N^3/2)$. The error in the Ewald summation method strongly depends on three parameters: the width of the Gaussians α, and the cutoffs of the real and Fourier space [49].

8.3.3 Other methods

Apart from the Ewald summation scheme, other methods have been developed, such as the $\mathcal{O}(N)$ fast multipole method [50, 51]. This scheme groups particles at a large distance into a cluster, for which it is not necessary to separately compute all particle–particle interactions. The charge distribution in such a cluster is typically approximated by a multipole expansion [41]. Other fast and efficient methods, such as **MMM2D** [52] rapidly converge for two-dimensional geometries and transform the Coulomb sum via a convergence factor into a series of fast decaying functions which can be easily evaluated, so that the scaling reduces even further to $\mathcal{O}(N^{5/3})$, with N the number of charges in the system.

The particle mesh methods, which are particle based, relate to the system of particles and convert this into a mesh of density values. The discretized Poisson equation is then solved on this grid. Such schemes are particularly of interest when a fixed cutoff is used for both the Fourier and the real part (short-range interactions) of the Ewald summation. In such cases the scaling of the method increases to $\mathcal{O}(N^2)$. The original **particle–particle particle–mesh (P^3m)** method [53] was based on an interpolation of the charges on the mesh points and on splitting the Coulomb potential into two parts in a similar manner as in the Ewald sum. The method can easily solve for the Poisson equation reducing the scaling to $\mathcal{O}(N \log N)$ due to the use of the fast Fourier transform (FFT) scheme [54]. N in this case is the number of points of the Fourier transformation. In MD simulations the forces acting on the particles are calculated through the electrostatic energy obtained from the solution of the discretized Poisson equation. In this force calculation, errors are introduced due to the interpolation of the particles on the grid, as the particles are forced to have a lower spatial resolution. The P^3m scheme overcomes this issue by calculating the electrostatic potential through directly summing the particles closely enough and using the particle mesh method for particles being further apart. Complementary schemes are the **particle mesh Ewald (PME)** [55] and the **smooth particle mesh Ewald algorithm (SPME)** [56]. PME splits the direct summation of interaction energies between point particles into a direct sum for the short-ranged potential in real space and a summation in Fourier space of the long-ranged part. The former is the particle part of PME and the latter is the Ewald part. SPME uses the smooth charge assignment functions involving (complex) B-splines when interpolating the charges on a grid together with an appropriately adjusted Green's function like in P^3m.

The choice of the method in electrostatic calculations depends on the systems and properties of interest. There is no exact recipe. For example, PME is better directed to infinite systems, which require the periodic symmetry. In systems with smooth variations in the charge density or continuous potential functions, the particle mesh methods are more efficient as the density field is restricted to a grid. On the other hand, schemes such as the MMM2D algorithm are fast and accurate in 2D slab geometries. The error estimation in these algorithms strongly depends on the Ewald parameters, which need to be chosen carefully. Compared to a standard Ewald sum, which uses the same number of k-vectors for the Fourier transform, all mesh algorithms are much less accurate. In practice, this does not cause any real problem, as typically a thermostat is used in the simulations at the same time as the electrostatic forces are calculated. The calculated forces are quite adequate with respect to the fluctuations the thermostat imposes on the system [55]. The schemes mentioned above can be extended or complemented by other more efficient ones depending on the application. The Ewald summation can be used also to deal with dipolar particles [57] or dielectric constants of polar fluids [58] and induced charges on arbitrary dielectric boundaries using the ICC* [59].

8.3.4 Electrokinetics

Efficient calculation of the long-range electrostatic interactions is essential in systems involving charges. It becomes more complicated when these charges move, for example in a flow. In these cases, electrokinetic effects are also important, they guide the dynamics of the system, and need to be treated in the simulations. These effects are responsible for electrophoretic and electro-osmotic phenomena [60–62]. In essence, electrokinetics involve electrostatic interactions between charged particles, but also involve the motion of these particles in a flow field [63]. These effects are particularly of interest in microfluidics [64] or soft matter and biophysics, which deal with electrophoretic phenomena, the behavior of colloids [65] or biomolecules in a flow [66, 67], various types of solutions [68], ion transport [69], active swimmers [70], etc. Such phenomena typically involve various time and length scales. For example, in complex fluids, solvent molecules with sizes of a few angstroms move along with suspended molecules, which can have a size up to hundreds of nanometers. Modeling these often needs coarse-graining approaches involving kinetic models or direct MC or MD simulations coupled to mesoscopic schemes. Typically, in order to include electrokinetic effects, the electrostatic potential is determined by means of the Poisson equation, the motion (including diffusion and advection) of the charges by the Poisson–Nernst–Planck equation [71] and the velocity for the fluid flow by the Navier–Stokes equation [72]. This can be done in a continuous space, as well as on a discrete grid, in which case these equations need to be discretized. Often a mean-field approximation for these phenomena is followed, which assumes a simplified representation of the physical system[2]. This is done with **finite-element schemes (FEM)**, typically used by engineers. FEM solves differential equations, like Navier–Stokes and Poisson–Boltzmann for discrete small parts of the system (the finite elements). For this, a very widely used sofware is COMSOL [73].

A number of different approaches have been developed for including an electrokinetic description in the simulations. In principle, these include the kinetic theory coupled with another relevant approach for monitoring electrokinetic phenomena. Examples are the Poisson–Boltzmann–Nernst–Planck model [74], which couples the Nernst–Planck equations with the Boltzmann distributions of ion concentrations. Most commonly mesoscopic models based on the lattice Boltzmann method including an electrohydrodynamic coupling can well capture electrokinetic phenomena in complex systems like colloidal suspensions [75]. Such mesoscopic approaches can also model electroviscous transport phenomena [76]. Electrohydrodynamics-based models are used to unravel the ion concentrations and electro-osmotic flow in a nanopore for DNA selectivity [77]. A microscopic self-consistent approach coupling the relevant equations of kinetic theory, classical density functional theory, and lattice-Boltzmann is capable of studying electro-osmotic flows under nanoconfinement or the modulation of the ionic current to DNA docking on a nanopore [78, 79]. Other numerical algorithms coupling the Poisson–Nernst–Planck equation for the electrostatic potential with the classical equilibrium density functional theory have been used before to model ion channels and extract the ion flux through these in view of biological ion-selective pores [80, 81]. On another level a combination of canonical ensemble MC and MD using PME has also proven quite efficient in investigating ion gating in water filled channels [82]. A combination of MD simulations with a mesoscopic description and a CG scheme

---

[2] As an example, in a mean-field representation a polymer or a biomolecule like DNA are modeled using stiff cylinders or rods.

can evaluate the ion dynamics in clay interlayers [83]. Note, that it is always important to be aware of the limitations of the theory, as well as the applicability of the simulations in order to apply these in studying electrokinetc phenomena [84]. Finally, the studies on electrokinetic phenomena in different systems extend way beyond what has been reviewed here.

8.4 Example codes

Moving towards the end, a short and only representative overview of simulation codes commonly used in physics will be given. These codes are academic or commercial software packages or can run under the GNU license [85]. They may include one or more of the simulation methods reviewed in this book. In the spirit of chapter [2], common DFT codes are SIESTA [86, 87], VASP [88], QUANTUM ESPRESSO [89], FHI-aims [90], ABINIT [91]. CP2K [92] and CPMD [93] can perform *ab initio* MD, also in the framework of the Car–Parrinello MD. More chemistry directed QM and DFT codes are Gaussian [94] or GAMESS [95]. Orca [96] includes a lot of capabilities for *ab initio*, DFT or semi-empirical calculations. Yambo [97] or DACAPO [98] are other *ab initio* codes. Some of these codes can also perform QM/MM simulations. Tight-binding schemes can be used under DFTB [99] or NRLTB [100]. The most commonly used classical MD codes are GROMACS [101], NAMD [102], AMBER [103], LAMMPS [104], CHARMM [105]. A very comprehensive list of QM and MD computational codes can be found elsewhere [106, 107]. CG and MS (e.g. by coupling MD with a mesoscopic solvent) can be provided through the use of VOTCA [108], ESPResSo [109], MUPHY [110], etc.

Typically for MS simulations homemade codes are also developed or utilities which can be added onto the existing academic/commercial codes. Specifically, for CG, the codes given above can be straight-forwardly used as long as a pre-processed coarse-graining has been done. Overall, in the distributions for the computational codes, often software patches or utility tools are provided to assist with the analysis of the results and their visualization. Finally, note that the codes are being maintained and new features are being added. In this respect, one should be aware that different distributions might affect the results and the same version of the code should be taken for a specific or a comparative study to guarantee consistency. In any case, the choice of the code is strictly related to the modeled system and the accuracy of the desired properties.

References

[1] Binder K and Heermann D 2010 *Monte Carlo Simulation in Statistical Physics: An Introduction* (Berlin: Springer)
[2] Debye P 1909 Näherungsformeln für die zylinderfunktionen für große werte des arguments und unbeschränkt veränderliche werte des index *Mathematische Annalen* **67** 535–58
[3] Hestenes M R and Stiefel E 1952 Methods of Conjugate Gradients for Solving Linear Systems *J. Res. Natl Bur. Stand.* **49** 409–36
[4] Hamming R 1973 *Numerical Methods for Scientists and Engineers* (New York: Dover)
[5] Scherer P O J 2015 *Computational Physics* (Berlin: Springer)

[6] Rapaport D C 2004 *The Art of Molecular Dynamics Simulation* (Cambridge: Cambridge University Press)

[7] Frenkel D and Smit B 2001 *Understanding Molecular Simulation: From Algorithms to Applications* vol 1 (San Diego, CA: Academic)

[8] Landau D P and Binder K 2014 *A Guide to Monte Carlo Simulations in Statistical Physics* (Cambridge: Cambridge University Press)

[9] Mouritsen O G 1984 *Computer Studies of Phase Transitions and Critical Phenomena* (Berlin: Springer)

[10] Kofke D A and Cummings P T 1997 Quantitative comparison and optimization of methods for evaluating the chemical potential by molecular simulation *Mol. Phys.* **92** 973–96

[11] Leach A R 2001 *Molecular Modelling: Principles and Applications* (Harlow: Pearson)

[12] Kofke D A 2005 Free energy methods in molecular simulation *Fluid Phase Equilib.* **228** 41–8

[13] Zwanzig R W 1954 High-temperature equation of state by a perturbation method. i. nonpolar gases *J. Chem. Phys.* **22** 1420–6

[14] Watanabe M and Reinhardt W P 1990 Direct dynamical calculation of entropy and free energy by adiabatic switching *Phys. Rev. Lett.* **65** 3301

[15] Kästner J 2011 Umbrella sampling *WIREs: Comput. Mol. Sci.* **1** 932–42

[16] Kumar S *et al* 1992 The weighted histogram analysis method for free-energy calculations on biomolecules. i. the method *J. Comput. Chem.* **13** 1011–21

[17] Laio A and Parrinello M 2002 Escaping free-energy minima *Proc. Natl Acad. Sci.* **99** 12562–6

[18] Laio A and Gervasio F L 2008 Metadynamics: a method to simulate rare events and reconstruct the free energy in biophysics, chemistry and material science *Rep. Prog. Phys.* **71** 126601

[19] Piana S and Laio A 2007 A bias-exchange approach to protein folding *J. Phys. Chem.* B **111** 4553–9

[20] Martoňák R *et al* 2005 Simulation of structural phase transitions by metadynamics *Zeitschrift für Kristallographie-Crystalline Materials* **220** 489–98

[21] Vargiu A V *et al* 2008 Dissociation of minor groove binders from dna: insights from metadynamics simulations *Nucleic Acids Res.* **36** 5910–21

[22] http://www.plumed.org/ (accessed 20 June 2016)

[23] Widom B 1963 Some topics in the theory of fluids *J. Chem. Phys.* **39** 2808–12

[24] Bennett C H 1976 Efficient estimation of free energy differences from Monte Carlo data *J. Comput. Phys.* **22** 245–68

[25] Voter A F 1997 A method for accelerating the molecular dynamics simulation of infrequent events *J. Chem. Phys.* **106** 4665–77

[26] Huber T, Torda A E and van Gunsteren W F 1994 Local elevation: a method for improving the searching properties of molecular dynamics simulation *J. Comput,-aided Mol. Des.* **8** 695–708

[27] Grubmüller H 1995 Predicting slow structural transitions in macromolecular systems: conformational flooding *Phys. Rev.* E **52** 2893

[28] Henkelman G, Uberuaga B P and Jónsson H 2000 A climbing image nudged elastic band method for finding saddle points and minimum energy paths *J. Chem. Phys.* **113** 9901–4

[29] Sprik M and Ciccotti G 1998 Free energy from constrained molecular dynamics *J. Chem. Phys.* **109** 7737–44

[30] Ferrenberg A M and Swendsen R H 1989 Optimized Monte Carlo data analysis *Phys. Rev. Lett.* **63** 1195

[31] Weinan E, Ren W and Vanden-Eijnden E 2002 String method for the study of rare events *Phys. Rev.* B **66** 052301

[32] Dellago C, Bolhuis P G, Csajka F S and Chandler D 1998 Transition path sampling and the calculation of rate constants *J. Chem. Phys.* **108** 1964–77

[33] Jarzynski C 1997 Nonequilibrium equality for free energy differences *Phys. Rev. Lett.* **78** 2690

[34] Jarzynski C 1997 Equilibrium free-energy differences from nonequilibrium measurements: a master-equation approach *Phys. Rev.* E **56** 5018

[35] Crooks G E 1998 Nonequilibrium measurements of free energy differences for microscopically reversible Markovian systems *J. Stat. Phys.* **90** 1481–7

[36] Crooks G E 2000 Path-ensemble averages in systems driven far from equilibrium *Phys. Rev.* E **61** 2361

[37] Steinbach P J and Brooks B R 1994 New spherical-cutoff methods for long-range forces in macromolecular simulation *J. Comput. Chem.* **15** 667–83

[38] Ewald P P 1921 Die berechnung optischer und elektrostatischer gitterpotentiale *Ann. Phys., Lpz.* **369** 253–87

[39] Cisneros G A, Karttunen M, Ren P and Sagui C 2013 Classical electrostatics for biomolecular simulations *Chem. Rev.* **114** 779–814

[40] Luty B A, Tironi I G and van Gunsteren W F 1995 Lattice-sum methods for calculating electrostatic interactions in molecular simulations *J. Chem. Phys.* **103** 3014–21

[41] Greengard L and Rokhlin V 1987 A fast algorithm for particle simulations *J. Comput. Phys.* **73** 325–48

[42] Sagui C and Darden T A 1999 Molecular dynamics simulations of biomolecules: long-range electrostatic effects *Ann. Rev. Biophys. Biomol. Struct.* **28** 155–79

[43] Pollock E and Glosli J 1996 Comments on P3M, FMM, and the Ewald method for large periodic coulombic systems *Comput. Phys. Commun.* **95** 93–110

[44] de Leeuw S W, Perram J W and Smith E R 1980 Simulation of electrostatic systems in periodic boundary conditions. I. Lattice sums and dielectric constants *Proc. R. Soc. London A* **373** 27–56

[45] De Leeuw S W, Perram J W and Smith E R 1980 Simulation of electrostatic systems in periodic boundary conditions. II. Equivalence of boundary conditions *Proc. R. Soc. London A.* **373** 57–66

[46] De Leeuw S W, Perram J W and Smith E R 1983 Simulation of electrostatic systems in periodic boundary conditions. III. Further theory and applications *Proc. R. Soc. London A* **388** 177–93

[47] Hansen J- P 1986 Molecular dynamics simulation of Coulomb systems in two and three dimensions *Proceedings of the International School of Physics Enrico Fermi Course* (Amsterdam: North Holland) pp 89–129

[48] Porto M 2000 Ewald summation of electrostatic interactions of systems with finite extent in two of three dimensions *J. Phys. A: Math. Gen.* **33** 6211

[49] Kolafa J and Perram J W 1992 Cutoff errors in the Ewald summation formulae for point charge systems *Mol. Simul.* **9** 351–68

[50] Appel A W 1985 An efficient program for many-body simulation *SIAM J. Sci. Stat. Comput.* **6** 85–103

[51] Barnes J and Hut P 1986 A hierarchical o(nlogn) force-calculation algorithm *Nature* **324** 446–9

[52] Arnold A and Holm C 2002 MMM2D: A fast and accurate summation method for electrostatic interactions in 2d slab geometries *Comput. Phys. Commun.* **148** 327–48

[53] Hockney R W and Eastwood J W 1988 *Computer Simulation Using Particles* (Boca Raton, FL: CRC Press)

[54] Press W H 2007 *Numerical Recipes 3rd edition: The Art of Scientific Computing* (Cambridge: Cambridge University Press)

[55] Deserno M and Holm C 1998 How to mesh up ewald sums. I. A theoretical and numerical comparison of various particle mesh routines *J. Chem. Phys.* **109** 7678–93

[56] Essmann U *et al* 1995 A smooth particle mesh Ewald method *J. Chem. Phys.* **103** 8577–93

[57] Toukmaji A, Sagui C, Board J and Darden T 2000 Efficient particle-mesh Ewald based approach to fixed and induced dipolar interactions *J. Chem. Phys.* **113** 10913–27

[58] Gray C G, Sainger Y S, Joslin C G, Cummings P T and Goldman S 1986 Computer simulation of dipolar fluids. dependence of the dielectric constant on system size: A comparative study of Ewald sum and reaction field approaches *J. Chem. Phys.* **85** 1502–4

[59] Tyagi S *et al* 2010 An iterative, fast, linear-scaling method for computing induced charges on arbitrary dielectric boundaries *J. Chem. Phys.* **132** 154112

[60] Sparreboom W, Van Den Berg A and Eijkel J C T 2010 Transport in nanofluidic systems: a review of theory and applications *New J. Phys.* **12** 015004

[61] Westermeier R 2016 *Electrophoresis in Practice: a Guide to Methods and Applications of DNA and Protein Separations* (New York: Wiley)

[62] Spanner D C 1975 Electroosmotic flow *In Transport in Plants I* (Berlin: Springer) pp 301–27

[63] Masliyah J H and Bhattacharjee S 2006 *Electrokinetic and Colloid transport Phenomena* (New York: Wiley)

[64] Li D 2004 *Electrokinetics in Microfluidics* vol 2 (San Diego, CA: Academic)

[65] Schmitz R, Starchenko V and Dünweg B 2013 Computer simulation of electrokinetics in colloidal systems *Eur. Phys. J. Spec. Top.* **222** 2873–80

[66] Fogolari F, Brigo A and Molinari H 2002 The Poisson–Boltzmann equation for biomolecular electrostatics: a tool for structural biology *J. Mol. Recogn.* **15** 377–92

[67] Lu B Z, Zhou Y C, Holst M J and McCammon J A 2008 Recent progress in numerical methods for the Poisson–Boltzmann equation in biophysical applications *Commun. Comput. Phys.* **3** 973–1009

[68] Fischer S, Naji A and Netz R R 2008 Salt-induced counterion-mobility anomaly in polyelectrolyte electrophoresis *Phys. Rev. Lett.* **101** 176103

[69] Pennathur S and Santiago J G 2005 Electrokinetic transport in nanochannels. 1. Theory *Anal. Chem.* **77** 6772–81

[70] de Graaf J *et al* 2016 Lattice-Boltzmann hydrodynamics of anisotropic active matter *J. Chem. Phys.* **144** 134106

[71] Zheng Q and Guo-Wei W 2011 Poisson–Boltzmann–Nernst–Planck model. *J. Chem. Phys.* **134** 194101

[72] Gross R J and Osterle J F 1968 Membrane transport characteristics of ultrafine capillaries *J. Chem. Phys.* **49** 228–34

[73] Pryor R W 2009 *Multiphysics Modeling Using COMSOL: a First Principles Approach* (Sudbury, MA: Jones & Bartlett Publishers)

[74] Zheng Q and Wei G- W 2011 Poisson–Boltzmann–Nernst–Planck model *J. Chem. Phys.* **134** 194101

[75] Pagonabarraga I, Capuani G and Frenkel D 2005 Mesoscopic lattice modeling of electro-kinetic phenomena *Comput. Phys. Commun.* **169** 192–6

[76] Warren P B 1997 Electroviscous transport problems via lattice-Boltzmann *Int. J. Mod. Phys.* C **8** 889–98

[77] Luan B and Stolovitzky G 2013 An electro-hydrodynamics-based model for the ionic conductivity of solid-state nanopores during DNA translocation *Nanotechnology* **24** 195702

[78] Melchionna S and Marconi U M B 2011 Electro-osmotic flows under nanoconfinement: a self-consistent approach *EPL (Europhysics Letters)* **95** 44002

[79] Chinappi M *et al* 2014 Modulation of current through a nanopore induced by a charged globule: Implications for DNA-docking *Europhys. Lett.* **108** 46002

[80] Gillespie D, Nonner W and Eisenberg R S 2002 Coupling Poisson–Boltzmann–Nernst–Planck and density functional theory to calculate ion flux *J. Phys.: Condens. Matter* **14** 12129

[81] Rosenfeld Y 1993 Free energy model for inhomogeneous fluid mixtures: Yukawa-charged hard spheres, general interactions, and plasmas *J. Chem. Phys.* **98** 8126–48

[82] Leung K 2008 Ion-dipole interactions are asymptotically unscreened by water in dipolar nanopores, yielding patterned ion distributions *J. Am. Chem. Soc.* **130** 1808–9

[83] Rotenberg B, Marry V, Dufrêche J- F, Giffaut E and Turq P 2007 A multiscale approach to ion diffusion in clays: building a two-state diffusion-reaction scheme from microscopic dynamics *J. Colloid Interface Sci.* **309** 289–95

[84] Hribar B, Vlachy V, Bhuiyan L B and Outhwaite C W 2000 Ion distributions in a cylindrical capillary as seen by the modified Poisson–Boltzmann theory and Monte Carlo simulations *J. Phys. Chem.* B **104** 11522–7

[85] http://www.gnu.org/ (accessed 21 July 2016)

[86] Soler J *et al* 2002 The siesta method for ab initio order-n materials simulation *J. Phys.: Condens. Matter* **14** 2745

[87] http://departments.icmab.es/leem/siesta/ (accessed 15 July 2016)

[88] https://www.vasp.at/ (accessed 15 July 2016)

[89] http://www.quantum-espresso.org/ (accessed 15 July 2016)

[90] https://aimsclub.fhi-berlin.mpg.de/ (accessed 15 July 2016)

[91] http://www.abinit.org/ (accessed 15 July 2016)

[92] https://www.cp2k.org/ (accessed 15 July 2016)

[93] http://www.cpmd.org/ (accessed 15 July 2016)

[94] http://www.gaussian.com/ (accessed 15 July 2016)

[95] http://www.msg.chem.iastate.edu/gamess/ (accessed 15 July 2016)

[96] https://orcaforum.cec.mpg.de/ (accessed 15 July 2016)

[97] http://www.yambo-code.org/ (accessed 15 July 2016)

[98] https://wiki.fysik.dtu.dk/dacapo (accessed 15 July 2016)

[99] http://www.dftb-plus.info/ (accessed 15 July 2016)

[100] http://esd.spacs.gmu.edu/tb/TBMD/index.html (accessed 15 July 2016)

[101] http://www.gromacs.org/ (accessed 15 July 2016)

[102] http://www.ks.uiuc.edu/Research/namd/ (accessed 15 July 2016)

[103] http://ambermd.org/ (accessed 15 July 2016)

[104] http://lammps.sandia.gov/ (accessed 15 July 2016)

[105] http://www.charmm.org/ (accessed 15 July 2016)
[106] https://en.wikipedia.org/wiki/List_of_quantum_chemistry_and_solid-state_physics_software (accessed 15 July 2016)
[107] https://en.wikipedia.org/wiki/List_of_software_for_molecular_mechanics_modeling (accessed 15 July 2016)
[108] http://www.votca.org/ (accessed 15 July 2016)
[109] http://espressomd.org/ (accessed 15 July 2016)
[110] Bernaschi M, Melchionna S, Succi S, Fyta M, Kaxiras E and Sircar J K 2009 MUPHY: A parallel multi physics/scale code for high performance bio-fluidic simulations *Comput. Phys. Commun.* **180** 1495–502

www.ingramcontent.com/pod-product-compliance
Lightning Source LLC
Chambersburg PA
CBHW081543220326
41598CB00036B/6544